ALIEN PARADIGMS

Refrain from illusions, insist on work and not on words. Patiently seek divine and scientific truth.

-the last words of Maria Dmitrievna Mendeleeva (mother of Dmitri Mendeleev)

1.0 Introduction

1.1 *Seeing Strange things in the sky*

Since the beginning of our time, we humans have sometimes seen strange things in the skies that have not fit into the conventional understanding of the normal world. In prior eras, if someone saw a strange object in the sky that seemed to defy the intuitive understanding of gravity or move at unbelievable rates of speed, they might naturally conclude that the sight was directly caused by a supernatural being briefly intervening in the human world. Today, we may say the sight was caused by an alien being visiting earth—and not a supernatural being—if we can rule out other potential causes such as human error or delusion. The alien being is a new type of conceptual category in the field of human thought. It is the result of the revolution in planetary science, evolutionary theory, and space technology, which started in the 1940's and continues to this day.

Although the idea of alien beings existing in the universe today is a conventionally accepted concept, the specific possibility that some of these beings might be visiting earth right now is not conventionally accepted and remains too difficult for many to even contemplate. But why are many humans resistant to the possibility of this idea? And could it be the case that the strange things we see in the sky sometimes are in fact real aliens?

1.2 *The Copernican Dilemma*

Many people understandably resist the notion that aliens could be the causes of any phenomena witnessed in the human world; after all, aliens do not form a part of the conventional experience of human history and knowledge; nor, for that matter, are there any other terrestrial, non-human species that practice math, science, or the arts of manufacture; nor is there a clear record of such species existing in the past on earth. However, if we look into the history of human thought, we see that conventional conceptual restrictions of this nature are sometimes in error. Such was the case with the Copernican Revolution in the mid 16th and 17th centuries.

Copernicus was the first modern astronomer to provide a set of systematic calculations of celestial bodies based on the assumption that the earth moved around the sun in space. At the time, conventional thought was grounded in the work of Aristotle and held that the earth was at the center of the universe and all things moved around it. However, calculations based on the Aristotelian model were clunky, inaccurate, and required a number of *ad hoc* adjustments. Copernicus used a different assumption (that the earth moved around the sun) and conducted a long series of observations and calculations based on this assumption.

When the calculations based on the Copernican model were published, the astronomical community immediately recognized them as superior to the Aristotelian model. But strangely, they just couldn't admit that Copernicus was actually right to assume that the earth moved around the sun. This attitude is typified by the following remark made by the English astronomer and navigator Thomas Blundeville in 1594, almost 51 years after Copernicus' work was first published.

> Copernicus . . . affirmeth that the earth turneth about and that the sun standeth still in the midst of the heavens, by help of which false supposition he hath made truer demonstrations of the motions and revolutions of the celestial spheres, than ever were made before.[1]

Today, we "know" that this statement is wrong from the standpoint of 21st century paradigms of scientific knowledge; for when we look back over the vista of time at the particulars of the past, we will sometimes see a range of undisclosed errors that those living in prior eras could not. From our perspective, the remark from Blundeville that a "false supposition" provides a basis for "truer demonstrations" makes no sense. How could a false supposition lead to truth? Shouldn't a false supposition always lead to error? What we see in this statement from Blundeville is an expression of a paradigm clash. This is a situation in which the paradigms upon which a group of people rely to conventionally reckon their senses in the real world clash with deeper truth that is emerging, and which is alien to that conventional paradigm. This is a process that has happened repeatedly in different fields of human thought over the past 500 years as discussed in this work. In these types of cases,

---

[1] A partial table of citations is at the end of this text.

the conventional human mind will want to reject this new truth or deny its reality. But the record of modern human achievement shows that strange new truth can emerge from these contradictions, if we are willing to push through the noise and discover a new signal.

There were many people in Copernicus' time who could not believe that the earth moved around the sun, even though real facts (i.e., the results of Copernicus' calculations) were telling them this was the case. This was due to the fact that the Copernican Paradigm was too alien to accept. A similar thing is happening today with what we will refer to in this work as "alien anomalies". An alien anomaly is an anomalous occurrence or event which is strange, bizarre, and unbelievable, and seems like the type of thing that an extraterrestrial alien would do. On one level, it may sometime seem like there is an actual alien doing something in the human world. But on the other hand, the idea of an actual alien intruding into human reality also seems like an idea from a science fiction movie, something which couldn't be happening in real life. This is our 21st century version of the Copernican dilemma.

1.3 *The Copernican Revision*

Another signal of this Copernican paradigm clash is seen in the actions of the Lutheran theologian Andreas Osiander who published a version of Copernicus' *De Revolutionibus* and added an unauthorized introduction that suggested the reader was at liberty to treat the earth's orbit around the sun as a mathematical fiction which should be used only for purposes of the calculations and be disregarded as corresponding to a physical reality. "Using Copernicus' mathematical system without advocating the physical motion of the earth provided a convenient escape from the dilemma posed by the contrasting celestial harmonies and territorial discord of the *De Revolutionibus*," the historian and philosopher of science Thomas Kuhn tells us. "It also gradually tempered the astronomer's initial conviction that the earth's motion was absurd."

People of the 16th century had very sensible reasons to think that the ideas of Copernicus were insane, for Copernicus' ideas were in direct conflict with the conventional understanding of how the world itself operated as well as *the experience of the world itself.* An example of these reasons was stated in an immensely popular French poem from 1578 (also published in England) as follows,

Those clerks who think (how absurd a jest)
That neither heav'ns nor stars do turn at all,
Nor dance about this great round earthly ball;
But th'earth itself, this massy globe of ours,
Turns around-about once every twelve-twice hours:
And we resemble land-bred novices
New brought aboard to venture on the seas;
Who, at first launching from the shore, suppose
The ship stands still, and that the ground it goes . . .
So, never should an arrow, shot upright,
In the same place upon the shooter light;
But would do, rather, as, at sea, a stone
Aboard a ship upward uprightly thrown;
Which not within-board falls, but in the flood
Astern the ship, if so the wind be good.
So should the fowls that take their nimble flight
From western marches towards morning's light; . . .
And bullets thundered from the cannon's throat
(Whose roaring downs the heav'nly thunder's note)
Should seem recoil: since the quick career,
That our round earth should daily gallop here,
Must needs exceed a hundred-fold, for swift;
Bird, bullets, winds; their wings, their force, their drift.
   Arm'd with these reasons, 'twere superfluous.
T'assail the reasons of Copernicus;
Who, to save better of the stars th'apperance,
Unto the earth a three-fold motion warrants.

This poem expresses the tension and confusion which arises when conventional understanding simply cannot make sense of a new set of facts about the world. The conventional reaction is to resist the new facts as being absurd and to stop considering them. An alternative approach is to work on overcoming this obstacle to arrive at a new vantage point from which to rationally view the new terrain. In order to get to this new place of understanding, it is necessary to ignore such feelings of absurdity and contradiction and press on in the field of thought. This is at least the example presented to us by the Copernican Revision.

A more scientifically expressed formulation of the same conceptual struggle against Copernicus was voiced by the 16th century philosopher Jean Bodin as follows,

> No one in his sense, or imbued with the slightest knowledge of physics, will ever think that the earth, heavy and unwieldy from its own weight and mass, staggers up and down around its own center and that of the sun; for at the slightest jar of the earth, we would see cities and fortresses, towns and mountains thrown down. A certain courier Aulicus, when some astrologer in court was upholding Copernicus' idea before Duke Albert of Prussia, turning to the servant who was pouring the Falernian, said: "Take care that the flagon is not spilled." For if the earth were to be moved, neither an arrow shot straight up, nor a stone dropped from the top of a tower would fall perpendicularly, but either ahead or behind. . . . Lastly, all things on finding places suitable to their natures remain there, as Aristotle writes. Since therefore the earth has been allotted a place fitting its nature, it cannot be whirled around by other motion than its own.

In addition to these reasonable concerns, the earth's movement in space also threatened to overturn the prevailing theological and metaphysical narratives of the day without offering any valid replacements. This is yet another level at which conventional knowledge can resist the consideration and assimilation of new facts. For example, there is quote attributed to Martin Luther that reads as follows,

> People gave ear to an upstart astrologer who strove to show that the earth revolves, not the heavens or the firmament, the sun and the moon . . . This fool wishes to reverse the entire science of astronomy; but sacred Scripture tells us [Joshua 10:13] that God commanded the sun to stand still, and not the earth.

"Copernicanism," Kuhn tells us, "was potentially destructive of an entire fabric of thought. . . . More than a picture of the universe and more than a few lines of Scripture were at stake. The drama of Christian life and the morality that had been made dependent upon it would not readily adapt to a universe in which the earth was just one of a number of planets. Cosmology, morality, and theology had long been interwoven in the traditional fabric of Christian thought described by Dante at the beginning of the fourteenth century. The vigor and venom displayed at the height of the

Copernican controversy, three centuries later, testifies to the strength and vitality of the tradition."

The earth-centered universe was embedded in many different narratives upon which the people of Copernicus' time relied for their personal meaning and sense of home. Accepting that the earth moved in space was an unsettling idea which aroused many strong emotions, causing severe cognitive dissonance at the start of that new era in thought. While one can empathize with the difficult position people of Copernicus' time were in with respect to his new ideas, they appear mad today. Copernicus appears to be the sane one, but in his time he was not viewed in that way at all. In fact, he was the mad one whom everyone made fun of. The same could be true today with respect to the alien question. In the future, it may be the case that people who come to truly understand the strange, emerging, alien phenomena that are coming into our world today, and have somehow brought them into conventional thought, may look back at those who are confused about today's phenomenon in a similar way and view them as being ignorant.

One can see how the strange contradictions on today's alien question are similar in certain respects to what the people in Copernicus' time were confronting on the "earth" question. Modern exoplanetary science, as well as the teachings of evolutionary theory, imply very strongly that intelligent beings with highly advanced technological capabilities likely exist on other planets throughout the universe; in other words, conventional thought accepts that aliens are real. At the same time, conventional thought ridicules and rejects those who suggest that aliens could be visiting earth right now. Even many scientists who are genuinely interested in this subject, simply refuse to admit out loud that some of what they are seeing may be caused by aliens. *See* **Sections 5.3.1** through **5.3.3** (discussing the uncanny fear and neurosis which sometimes arises when humans contemplate sharing the earth with aliens). These people are victims of the same absurdity trap that plagued people in the days of Copernicus; they knew that what Copernicus was saying about the earth was true, but just couldn't admit it to themselves and others, instead choosing to engage in insane rationalizations to avoid confronting the truth.

To be fair, however, if aliens are actually visiting earth, they are not making their presence known in an unmistakable, global, and irrevocable manner, or otherwise acting like one might expect them to. Even though some claim to have had contact with aliens, no one sees them as a part of daily human life in the same way that other people, animals, and other aspects of the natural world are seen. This situation makes no sense from the perspective of conventional thought; why, after all, would aliens only have some contact with some people, under strange, unusual, and difficult to detect conditions, and not be visible in the world at large in a normal way? Why wouldn't they make a grand arrival in the real world? *See* ***Sections 3.3.3*** (discussing the arbitrary aspects of the "SETI" paradigm).

Moreover, the only time that one sees alien to human contact in a way we would expect is in the genre of science fiction fantasy. Because of this conditioning and experience, many have instinctively and irrevocably concluded that actual contact between humans and aliens could only be the "stuff of science fiction" and will not consider that alien presence in and around earth today could be a real occurrence, unless it is seen in the same way it is portrayed in the science fiction fantasy genre.

This position, while conventionally appropriate, is nevertheless problematic. In science fiction fantasy we project to ourselves what we would like to see. But at the same time, any such expectations regarding how aliens would or would not behave, however, are simply subjective, local expectations. They are not based upon an objective, universal study of alien life (because such a study does not exist). Thus, while these suppositions may appear to be "rock solid" to a conventionally conditioned mind, they are actually only a starting point for the inquiry and nothing more than that. These local expectations should yield to whatever is shown by the objective nature of the phenomena themselves subject to the rigorous application of reason. *See* ***Section 5.3.5*** (discussing how we correct the barriers to proper paradigm formation in the context of alien anomalies).

Thus, if it truly appears that some alien intelligent entities are visiting earth's aerospace in an irregular and surreptitious way, then that is how the phenomena are being presented, even if they are not understood clearly and fully. If there is a disconnect between human expectations and what is actually witnessed, the accuracy of our presuppositions and the completeness of our paradigms is what should be questioned, instead of denying the alien reality. In this way, the strange and irrational presentation of an anomaly is simply construed as a signal that present knowledge and understanding are incomplete, and nothing more than that. This constitutes an invitation to revise present expectations and paradigms, as well as their associated activities, in order to apprehend clear knowledge. But unfortunately, this is not how things presently work out within the realm of conventional understanding. Instead of being curious, interested, or delighted that something new is making its arrival known in our world, the fearful simian mind chooses to look away and go back to reading out the world it thinks it already knows.

1.4 *The Copernican Injunction*

Contrary to the conventionally accepted assumptions, this work assumes the possibility of intelligent alien life as a real one, and accepts it is possible (or even probable) that aliens have visited earth in the past and may be visiting earth right now. This is what the Copernican Injunction demands—treat the absurd postulate as possibly real and do the best you can to work through it. This alien assumption, however, is only an untested hypothesis at this point, based primarily on anecdotal evidence with a dash of direct physical material, and may or may not be true at the end of the day, as there could be other, undisclosed causes for alien phenomena.

Copernicus did not arrive at his conclusions overnight. Instead, he explored an idea which he thought could be true for a long, long time before he published his findings. Although Copernicus' ideas clashed with the local human experience of his day, he was not content to simply rely upon what the limitations of conventional knowledge were telling him was possible. But the criticisms and concerns that animated Copernicus' detractors must have also affected him too. His findings of planetary motion did nothing to address those concerns at all and instead created more anomalies than it solved. Like his detractors, Copernicus must have also felt the absurdity that resulted when his ideas clashed with the rest of the human world. And yet, somehow he stayed the course through it all.

Even though Copernicus could see that the earth moved in space, he himself could not explain why buildings did not fall, water poured into flagons, and cannonballs struck their targets. Paradigms which explain such phenomena in the Copernican universe (such as acceleration, gravity, and centrifugal force) were not to come for centuries. Yet, he was able to accept the incompleteness and remain undeterred, even though he was not capable of developing a theoretical framework that would reconcile the dissonance created by his work and the real world in which he lived. The Copernican Injunction thus commands us to be leery of the effects that the weight of conventional knowledge may have on the mind and to disregard false feelings in favor of productive pathways of knowledge. It also reminds us that it takes time to work out all the details in cases of sufficiently revolutionary paradigm shifts.

1.5 *Stagnation in the domain of alien knowledge*

A comparison of the state of knowledge of alien anomalies today with what was known over 60 years ago shows little change. In any other realm of knowledge, this would be completely unacceptable, but in this field, it seems to be the accepted norm. Due to a variety of historical accidents, however, today it is now possible to find out whether we are being visited, whether we are being delusional, or something in between. But due to the inertia created by conventional thought itself, stagnation still dominates this domain.

Since the 1940's, humans have spent more and more time in the aerospace due to the increase in nautical and aviation technical capabilities in response to the demands of World War II. Additionally, humans have developed technologies which allow for an observational awareness of the aerospace with great precision and depth. On top of that, conventional thought itself says that aliens are probably real. These convergent factors now make it possible to make real increases in our knowledge of these strange aerospace phenomena. But for reasons which are further explored and analyzed in this text, this has not yet happened. In order to change this state of affairs, the patterns of thought which give rise to this stagnation must first be analyzed and then reconsidered. In the negative space created by this rational critique, new forms of knowledge may one day emerge. *See* ***Section 5.3.1*** (identifying the neurotic side effects of alien anomalies).

Now, to be clear, the subject to alien anomalies is certainly getting more real attention than it ever has, which is a precursor perhaps to a more serious investment of time and effort. But collective interest is not the same thing as real knowledge. One reason for the continued lack of progress appears attributable to an absence of an accurate collective understanding of how scientific process can break this logjam and show us how to approach these alien anomalies in a "Copernican" way.

Increasing awareness of the operation of conventional thought, and applying this enhanced awareness to the subject to alien anomalies within the context of scientific understanding, specifically demonstrates that the human reaction to these alien anomalies has been stymied by science fiction conditioning, as well as the false belief that we have a comprehensive understanding of how the universe works and are just filling in the final details. When this is clearly seen, a better path reveals itself. To look at this from another angle, one could say that if intelligent aliens are really visiting earth from other places, they are not just bringing new gadgets or technologies, but rather a radically different paradigm of the cosmos, one which is alien to human understanding. The application of reason to these issues makes it possible to sense the presence of an alien paradigm just over the horizon, to see through limiting delusions, even if we cannot fully articulate this paradigm yet. To do this, we must learn to restrain ourselves from engaging in unproductive and excessive speculations which may feel good but ultimately teach nothing new at all. *See* ***Section 5.3.5*** (noting that the possibility of human to alien contact seems to drive some people a little mad and suggesting that this over-excitement is anti-scientific).

In this connection, we must acknowledge that modern conventional thought has yet to grasp the fact that it is completely unprepared for alien anomalies because of the lack of paradigms which might describe them and give them real meaning in the conventional world. This is just like the situation faced by Copernicus when he knew his core paradigm was right, but had no way to explain conventional phenomena such as gravity, acceleration and the like, since these paradigms would not arrive for centuries.

Accordingly, we are forced to conclude that the transformation of alien anomalies into conventional knowledge will only come through a scientific approach to this subject. This does not mean that we simply add "new data" to present understanding, but rather drastically revise how we think about this subject *before and as* we approach it. No human alive today has ever confronted a paradigm shift of this magnitude. This is why we must look back to the Copernican Revision as a guide and also delve deeply into the nature of paradigm shifts and scientific thinking itself, instead of engaging in gross speculation and useless sensationalism about this subject. Today's stagnation can only be overcome by drawing on the Copernican Revision itself and heeding its lessons. In this connection, *Alien Paradigms* is primarily focused on realigning these thought patterns so that you can see this for yourself.

1.6 *The value of an alien study*

While some might think it rather impractical, or perhaps foolish, to even consider investing real time and real money into studying these strange, alien anomalies, from a certain point of view one could say that a serious investment of study into this realm is quite practical. History has taught us countless times that domains of knowledge that initially seem impractical, or of only marginal value, can turn out to be immensely critical later on. Many ideas that get initially laughed off stage (such as the modern conception of paper money and credit, or the concept of human flight) turn out to constitute core features of a human world that people can't even imagine living without once they become real. Indeed, the Copernican Revision itself stands out as a prime example of such a thing. Many modern scientists believe that the Copernican Revision set off a conceptual chain reaction that made all of the modern sciences—and their products—possible. But at the time, Copernicus' ideas seemed annoying, trivial, and completely useless. People of the 16th Century certainly had much better things to do with their time than waste it on trying to figure out goofy, irrelevant details such as whether the sun revolved around the earth or not; after all, how could such knowledge ever be useful in the "real world" of human affairs? And yet, there could be no way that these people could have foreseen how the work of Copernicus would actually change the world.

Indeed, one could argue that a precise and exacting understanding of the nature and origin of some, if not all, of these aerospace anomalies could lead to a drastic revision of our understanding of the possibilities of science, and of reality itself. *See* ***Section 5.4*** (on the formation of alien paradigms). Even if none of the anomalies that humans witness or experience are caused by aliens, they are certainly caused by something; and whatever that something is, it is certainly "alien" to our world, coming from somewhere outside of the realm of our present understanding, *yet within the realm of our real possibilities.*

1.7 *Modern scientific vision offers new ways to look at alien anomalies*

As will be examined in greater detail below, none of the prime categories of our normal knowledge regarding the causation of phenomena can adequately explain all of these anomalies; and yet, at least some of these alien anomalies appear to be as real as any other phenomenon experienced in the world. This causes great dissonance because the point of having prime categories of causation is to help us understand and explain things. Only when we have categories of knowledge within which these anomalies can be described and studied, may we have new forms of conventional knowledge which operate in the normal human world. Until that time, we have no such knowledge and no conventional way to discuss and work with these phenomena, and the phenomena themselves appear "unreal" to the human mind. *See **Section 2.9*** (the real, the unreal, and the "Copernican Trap"). By acknowledging that such knowledge is within the realm of real possibilities, but outside the realm of present understanding, we recognize that our experience of this aspect of reality (i.e., that of intelligent extraterrestrial aliens) has not yet caught up to what reason tells us is likely true. This is analogous to what occurred during the Copernican Revision and can be addressed through scientific understanding over time.

Today, we are guided by a scientific understanding that connects human reality to universal reality in a new kind of way. In this new scientific vision, the whole of human reality, including all of our thoughts, ideas, and history, can be viewed as nothing more than one local example of a particular type of intelligent being. The whole of human reality, and all that can be perceived and experienced by humans, is simply one type of world. We can no longer say that our world is "the world". We can longer place special emphasis on the features of the human species or reasonably claim that a "Universal Creator" has chosen humans as the universe's most important or central species. We also can no longer hold fast to the notion that humans were created in the exact image of the Universal Creator, because we now must also concede that other intelligent beings, with highly advanced understandings of the laws of physics and science, likely exist alongside us in the universe as well. These beings, science implies, are not gods, but rather real beings, similar to us in some ways perhaps, but also vastly different in ways we cannot imagine. How, precisely, these beings might differ from, or be similar to, our own species is one of the questions that cannot be resolved through speculation, but rather only through factual inquiry.

Because these modern scientific ideas are floating around in our minds today, we can see things in the skies far differently than was the case in the past. When our ancestors may have seen strange things in the skies and interpreted them as gods, angels, or other supernatural beings, they were reflecting back what they expected to see within the limits of the frames of knowledge they were operating in. We today can say that some of things they saw could have been the caused by real beings, not mythological entities, and now understand them in a manner that was not even possible in prior eras of human thought. Because they are not gods to us, we can investigate them and develop new details about how they operate. As a result, there is new ground to explore and map out that was simply not available for inspection in prior eras.

We also have a technological capability to truly and accurately document a wide range of physical occurrences in our aerospace in a manner that was never possible before. If we wanted to, we could have complete visibility into nearly every type of object that is presently passing through our aerospace; and, indeed, the various militaries of the world do possess and deploy such a capacity right now. But because these capacities are completely dedicated to the maintenance of national security, and not general human knowledge, whatever knowledge is being gained from these sources has not been disclosed within the public record. This is something we will want to change, of course.

Based on the prior experience of the Copernican Revolution, we can confidently state that having some type of understanding of these anomalies—instead of always saying that they are "unidentified"—will teach us something new about the world we inhabit. We may not know what this new something will be yet, but that part doesn't matter right now. The important point is that there is something occurring in our near-earth environment which we do not understand, and it is now within our capability to at least learn something more about it if we have the open-mindedness and willingness to do so. Even if that something isn't itself much, the very act of trying to learn about these anomalies will teach us something new because it will force us to make a *new kind of scientific effort*.

For reasons explained in greater detail below, we can confidentially say today that it is worth placing the same level of active attention, investigation, and resource commitment into examining the real time occurrences of such aerospace anomalies as we would with the other normal objects of science such as planetary bodies, subatomic particles, and other living beings. But instead of only looking for anomalies at impossibly far distances, or fixating on anomalies *that have passed*, we must study anomalies *as they occur*, gathering active data about them, similar to the ways in which firefighters respond to fires while they are burning instead of waiting to show up until everything has been turned to ashes. We must develop new forms of active inquiry, similar to the forms of active inquiry used in other fields, such as nuclear science or biology, instead of simply remaining content to engage in speculation or abstract conceptual theorizing in the absence of any forms of conventional evidence.

One way to embark on this kind of approach, is to reexamine how we think and speak of these anomalies from the beginning; for if our thoughts are filled with error and delusions, so will be the results of our conduct. For example, most of the investigations in this domain are dominated by questions such as, "Are we alone?" or "Why aren't aliens here?" But, as will be demonstrated below, these questions contain a number of misleading embedded assumptions that always lead us astray. The only question that we should be asking of a given anomaly is: "what on earth is *that*?" This type of question signals the beginning of a clinical examination, similar to the ways in which a modern physician investigates an anomalous presentation of disease in the human body. In the clinical field, disease is studied in real time. In cases of an unknown presentation of disease, a physician doesn't endlessly question whether the disease is real but simply acknowledges that she doesn't yet know what it is and uses her lack of knowledge as a starting point for the inquiry instead of making no real effort to learn anything new. Clinical medicine would have never made any significant advances if clinicians had been content to simply shrug their shoulders at the new presentation of disease and say "unknown" or "unidentified". This provides a good model to follow with respect to alien anomalies today.

In this connection, we must examine how much collective time and resources are expended on activities which have no reasonable prospect of generating real knowledge but only stir up emotionalism and excitement. There is enough collective attention on this subject as it is, but the continued obsession on "disclosure" and making the subject of aliens mystifying, spooky, and weird, seems foolish now because it does not lead to any real knowledge and only serves to stir up strange emotions and religious mystery. Why continue doing such things that lead to no new results? Why continue to remain obsessed with revelatory "disclosure" when it never fails to disappoint?

Similarly, we should strongly consider ridding ourselves of the childlike demands for "extraordinary proof" of alien anomalies. Such demands are immature and completely inconsistent with the historical practice of normal scientific development. Those types of demands now need to be abandoned in lieu of an approach which is more consistent with the history and tradition of modern scientific advancement itself. This too is discussed in greater detail below.

Along the same lines, we must end the conventional practice of referring to these anomalies with the predicate "unknown" or "unidentified", as the unabated use of such conventional terminology continually suggests to the human mind that there is "no cause" for the anomaly in an unproductive way. The relentless suggestion that these anomalies cannot be identified is creating a collective "doom loop" which inhibits the development of a positive frame of reference within which these anomalies can be understood and worked with, since a network of use and meaning cannot be created around "no thing". Put differently, so long as language suggests to the human mind that there is "nothing really there" that can be understood, it is extremely difficult, if not impossible, for a real form of momentum to arise which might one day render these anomalies into ordinary occurrences within the world of human thought, language, and experience. As will be demonstrated below, scientific process can address and resolve this type of dilemma.

1.8 *The goal of studying alien anomalies*

As discussed below, the first serious scientific foray into the field of alien anomalies was presented in the work of Cocconi and Morrison in 1959. *See* **Section 3.3.1** ("aliens as inferred objects"). The initial goal of this study was to establish a "channel of communication" between humans and alien species. *See* **Section 7.4** ("Cocconi and Morrison's revolutionary vision"). This still should be the goal, but today is approached quite differently. In Cocconi and Morrison's time, the idea was that such a channel could be established through the transmission and reception of deep space radio signals from aliens. This no longer makes any sense. What does make sense, however, is to see if we can establish channels of communication with aliens that are actually visiting earth.

On the surface, this absurd idea sounds quite simple; after all, if aliens are here visiting earth, surely they would be eager and willing to communicate with us. The reality, however, is far stranger. Based on the scant evidence we have, it does appear likely that some type of alien presence is visiting earth. But we have no idea "what" this alien presence is, "when" it will visit, "why" it is visiting, and "how" we might communicate with it. It is a very frustrating and confusing scenario. Unravelling this bizarre situation, and understanding how we might make some kind of progress, is the primary purpose of this work which starts at square zero in the field of thought and moves forward from there.

1.9 *Summary of the approach*

This work presents itself in the format of a conventional philosophical argument, with appropriate references to various authorities in order to appeal to the reader's sense of reason. The presentation of the argument can be summarized as follows.

Section 2.0 entitled "Alien Anomalies", attempts to explicate the basic categories that we use in regular thought to explain occurrences in our world, and shows the reader that a different way to speak and think about these anomalies must now be adopted.

Section 3.0, "Safe Distance Paradigm," analyzes and criticizes the conventional scientific approach to alien anomalies within the general structure of scientific processes and paradigm shifts (i.e., the "search for extraterrestrial life," or "SETI") and shows how it is founded on a number of arbitrary and irrational suppositions.

Section 4.0, "Scientific Conversion Problems," directly addresses the issue of scientific revolutions, paradigm shifts, and the scientific conversion process, in the context of alien anomalies.

Section 5.0, "Alien paradigms," attempts to demonstrate how a paradigm around alien anomalies might actually be constructed within the field of thought.

"Alien experiences," Section 6.0, looks at alien anomalies from the human side and calibrates the weight to be given to reports of alien experiences within the overall context of scientific process.

Section 7.0, "Conclusion," attempts to tie the loose threads together and makes suggestions as to various lessons learned from this work and productive pathways we can pursue in the future.

2.0 ALIEN ANOMALIES

2.1 *Overview of the prime categories of conventional thought*

Today's conventional fund of human knowledge does not admit or accept that an extraterrestrial alien presence could be an actual cause for any type of phenomena that is experienced in the real world. Instead, the conventional fund allows only two general types of ultimate causes for any perceived phenomenon: human or natural.[2]

Natural causes include those types of phenomena which result from the actions of non-human terrestrial beings (e.g., animals, plants, insects, cellular entities, etc.), or forces of nature (e.g., volcanoes, storms, tidal forces, gravity, electromagnetism, etc.). We generally refer to scientific disciplines which study these phenomena as biology (the "study of life") and physics ("the study of natural things").

When examining phenomena, we seem to discriminate between "living beings" and "non-living things" automatically based on an implicit distinction we draw based on "the self" and the world "outside" the self. While we don't know for certain how this state of affairs arose, and we ourselves certainly did not create it, we accept it as the basis for conventional systems of thought, perception, and experience. As described by the late physicist David Bohm in *On Creativity* (1996),

> [F]rom early childhood we learn to accept the notion that the world is constituted out of a tremendous number of different and separately existent things. Some of these things are inanimate objects, some are alive, some are human beings. And to each person there is a certain very special one of these things, which is *himself*. This "self" is viewed, in the first

[2] When we define an "ultimate cause", the phenomenon has been placed in a settled order, allowing us to either move on to something else, or work with it in a relatable way if we chose to. Phenomena whose ultimate causes are "unknown", are either ignored, or given a spooky, supernatural, or "out of this world" significance. Ultimate causes are those which are firmly embedded within the world of human time (which begins in life and is timed out at death).

instance, as a physical body, sharply bounded by the surface of the skin, and then as a "mental entity" (also called the psyche or "the soul") which is "within" this physical body and which is taken to be as the very essence of the individual human being. The notion of a separately existent "self" thus follows as an aspect of the generally accepted metaphysics, which implies that *everything* is of this nature.

From this starting point of the self, things seem to go forward for us in the relative world of time, space, and matter. We see living beings as having a spatial intimacy, presence, or form of motion which seems to be caused by forces such as instinct, intention, or intelligence in a way that is somewhat familiar to our own selves. This perception then forms its own type of order. Non-living things, by contrast, appear to arise within a different order and move according to a "mechanistic" framework, as described by the laws of physics. Whether beings take the form of an insect, plant, animal, bacteria, or other type of cellular entity, we feel that all beings are part of the common order of life, which is objectively separated from the inert, non-living order. Natural things, such as rocks, volcanoes, planets, comets, lightning strikes, and the like, are generally viewed by us as moving due to forces such as gravity, the strong/weak nuclear forces, or electromagnetism. We generally do not attribute intention, instinct, intelligence, or any type of "personality" to natural or inert objects, nor do we consider them to be "alive". Instead, we observe natural phenomena to look for impersonal "laws" which explain their presentation and appearance.

With that said, even within the examples listed above, sometimes the line between living and non-living becomes blurred for us. For instance, some people view planets as living beings, as opposed to inert objects, and at least have an argument to make based upon how we chose to define the term "living". In *Finding the Mother Tree: Discovering the Wisdom of the Forest* (2021), Professor Suzanne Simard makes a compelling argument that forests participate in a form of sentient conduct. Based on her initial anecdotal experiences as a child growing up in the wilds of British Columbia, her subsequent work as a forester, and her later work as a research scientist, Simard describes the ways in which natural entities in a forest seem to communicate and work together in ways that, for lack of a better word, appear to be deliberate and teleological in nature. Simard's work plausibly demonstrates that within the world of the forest there exist systems of communications among natural entities, "paradigms" one might say, that allow the forest to work together as a whole and adapt to changing conditions over time in a way that would not be possible if there was no overall system of connection and communication. Simard's research blurs the bright lines that humans conventionally draw between the form of awareness they are directly cognizant of and the so-called inferior forms of awareness that they may believe guide the life of natural entities such as trees and plants in a more mechanistic, and thoughtless way. Thus, based on this approach, one could say that the earth is alive and doing things.

Similarly, some might say that a genetic entity straddles the line between living and non-living due to its direct relationship with our present understanding of the basis for life. On the one hand, genes are viewed as impersonal forces which determine the characteristics of living beings and their behaviors. Genes themselves don't have "personality" but many believe that genes play a major role in the display of personality by various creatures (personalities are entities that are alive and active for us and are fellow possessors of "living selves" and with which we can interact in a direct and unmediated way). On the other hand, genes are assumed to be part of the life system environment based upon the circumstances under which genetic material was first discovered, the effects that genes seem to have on living organisms, and the evidence we have accumulated from our study of terrestrial evolutionary biology.

The two prime categories of ultimate causes—human or natural—express a fundamental truth about how we see the world, what we think the world is ultimately made of, and how we make sense of the world today. These prime categories are the product of our collective history, experience and knowledge. They are expressions of how we ourselves see the universe, but they are not expression of how the universe actually is, any more than expressions of the universe were before the Copernican Revision changed them.

2.2. *The "alien" case of the beaver*

The case of the beaver[3] illustrates the ways in which the lines between impersonal natural forces (i.e., "genes"), and the active forces of life can become blurred. It is instructive to think about beavers in relation to the possibilities of extraterrestrial aliens because beavers themselves display characteristics which are alien in the sense that they defy expectations we might have about lower animals, including the assumption that such "lower" creatures lack an active, spontaneous, intention that humans typically associate with themselves. Beavers do things that humans may not expect and show an independent initiative; aliens, we must assume, also possess such characteristics. Unlike aliens, however, whose doings could be inherently baffling, incomprehensible, or even magical to us, *see* ***Section 4.8*** (discussing Arthur C. Clarke's dictum that says "any sufficiently advanced technology is indistinguishable from magic" in the context of the scientific conversion of alien anomalies), we can fully understand the precise scope and nature of beaver doings even if we cannot get inside the beavers' minds and have a direct discussion with them about the internal process they use to direct those doings. We can observe the actions of beavers from the outside, look at what they are doing, think about our own situation, and make inferences about the internal states of the beavers themselves. No scientist ever bases their knowledge of the beavers on the notion that the beaver is directly talking to the scientist and describing the internal states and processes that the beaver goes through in order to make its decisions. From the conventional perspective, we would think someone mad if they told us they were doing such a thing.

By contrast, when it comes to aliens, there is a presumption that we could engage in direct communication with them because they occupy an order of thought that is greater than our own. We would naturally assume that if they can develop powerful scientific technologies, they must know how to speak and communicate at a level which is on par with human thought; we do not expect the same, however, from beavers. But what do we do if we are prohibited from having the same type of communications with aliens like we would do with other humans? Does this mean we

---

[3] An examination of individual cases is useful to show what kinds of general principles may be in operation in our world because individual cases can only arise when certain types of conditions make them possible.

are completely prohibited from developing any real knowledge about their internal states?

If we witness or observe an anomalous phenomena which appears technical, unnatural, and could be the result of an alien force, and we are unable to direct communicate with the alien about what they are doing, then our ability to study aliens would be structured (and limited) in the same kind of way that our study of beavers would be. In other words, we do not have to wait until we are in a position to get a direct communication from an alien to learn something about the status of their internal selves, just like we do not have to wait until we can talk directly to the beavers in order to learn something about their internal states.[4] Thus, thinking about our attitudes towards beavers, and how we can draw inferences about their internal processes without having "direct access" to them through direct "conversations" (like we might do with other humans), gives us a basis from which we might draw analogies when determining how to think about aliens, as will be demonstrated later in this work.

Now to the beaver. The beaver is one of those rare creatures who, like human beings, engages in engineering activities which alter the mechanics of their natural environment to serve their needs for food and shelter. As discussed by Professor Donald Griffin in *Animal Minds: Beyond Cognition to Consciousness* (2001), they fall trees, construct lodges, dig burrows, create channels of water, make islands, and engage in dam building behavior to produce ponds. Not only do they engage in these types of construction activities, but they are also very active in maintaining these structures under changing and unpredictable conditions and are able to adapt spontaneously to alien circumstances with complex behavioral responses. In one case, for example, scientists put beavers in an artificial enclosure where a pipe fed water to a pool. The beavers decided they wanted to regulate the flow of water and quickly learned to plug the pipes with "calibrated sticks" whose ends were gnawed off to perfectly plug the diameter of the pipe. This type of situation was clearly unknown to the beavers before they were placed in it and could not have been something their ancestors were forced to confront. Yet, they were able to adapt "in

---

[4] With aliens, it is universally believed that they must hold some secret information and powerful knowledge and that we could gain access to this alien knowledge if we could communicate with them directly. Some people are so fixated on this that they won't even think of anything else being possible other than a direct alien transmission.

the moment," so to speak, thus demonstrating what appears to the human mind to be an active, spontaneous, and engaged intelligence.

The behaviors of beavers are accessible to direct observation over periods of stable time. This has allowed scientists to study the engineering decisions of beavers. As a result, they have learned that beavers take into account the characteristics of their environment when building their structures. For instance, when digging burrows and constructing lodges they appear to always chose locations where the shores have an appropriate degree of slope that is not too gradual. Beavers "explore shorelines and select for burrowing," Dr. Griffin tells us, "only places where the ground rises far enough to provide space for a dry chamber within a reasonable distance of the water." The selection and execution of building strategies suggests that the beavers have "ideas in mind" before they begin their actions. As described by Dr. Griffin,

> Aescbacher and Pilleri (1983) found that when captive beaver built a lodge they "cut sticks of two different lengths from the branches provided, peeled them and inserted the longer ones into the roof of the main chamber and the shorter ones into the chamber entrance, in a more or less radial pattern which was afterwards plastered with mud, wood shavings, and small twigs." Richard also describes cases where the beaver gathered appropriate material before starting to build, suggesting anticipation of the construction. Under natural conditions, beaver add to and alter lodges and burrows, so these shelters are in a constant state of enlargement and modification, somewhat like medieval cathedrals. Lodges and burrows are often abandoned, even when the beaver remain in the vicinity and build new shelters. Even after they are abandoned beaver lodges are substantial piles of material, sometimes 2-3 m high and 5-10 m long, and their remains can be recognized years later even though grass, bushes, and even trees may have grown up on them. Abandoned lodges are sometimes reoccupied and reconstructed.

Because some of the behaviors of beavers seem similar to our own ways of thinking and acting, we are naturally drawn to the question of whether the actions of the beavers are entirely the product of genetically determined instinct and response, or whether the beavers are displaying a type of intelligence that is not entirely determined by genetics or mere instinct and is on par with human thought. *See* ***Section 3.3.3.2.1.5*** ("The 'divine moment' error"). Since we do not have direct access to the "thoughts" of the beaver in the same way that we have access to the thoughts of other humans, our answers to these questions are solely derived from observing the externalized actions of the beavers and drawing the best inferences we can about what is going on inside of them.

Similarly, if aliens are real and if they are passing through or otherwise visiting the aerospace from time to time instead of sitting down with us and having direct conversations, we will not have reliable, objective access to what is going on inside of them in the same way that we might know what is going on inside another human being. Humans can ask each other why they are doing things, and what they are doing. Similarly, humans can ask questions of the world or universe and receive "answers" through scientific inquiry. But for reasons that no one understands, aliens are not communicating with human beings in a way that we would naturally expect a being of higher intelligence and technological power to do. We would expect that if a superior being came all the way to earth from some far flung region of the galaxy and saw us down here on earth, they would want to communicate and connect with us normally (this is the subconscious impulse which gives rise to the collective delusion known as "Fermi's paradox" as discussed in **Section 3.3.3.1**).

While there are some people who have claimed to have such alien contact, it is undisputed that no form of normalized, regular, and repeatable communication between aliens and humans has ever occurred.[5] So, for instance, even if some people say that they are in communication with aliens on a regular basis, no one has been able to bring another person into the conversation in the same way that we would do with other humans. Indeed, one can safely say that if a human being was ever able to do such a thing (i.e., bring any other human being into communication with alien beings in a way that was just as normal or conventional as the socialization that naturally occurs within the human social order itself) it would be a revolutionary occurrence of the most profound nature.

---

[5] Claims of personal "visionary" communications with aliens are discussed in Section 6.0 ("alien experiences").

The same is true, of course, with creatures like a beaver. No one knows how to sit down with a beaver and gain direct knowledge from them about how they construct their dwellings and the internal processes of deliberation that it appears they go through in order to arrive at their built results. We can see that some kind of process exists by studying beavers in a way that is similar to how we have learned to study ourselves. Of course, it would be amazing if we could talk with a beaver directly—and who knows, maybe one day we will know how to do those kinds of things. But the absence of this direct communication channel doesn't render us powerless to learn anything so long as we have sufficient time and space to observe the actions of beavers.

Biologists seek to understand beaver behaviors and they have enjoyed some success in this endeavor. Part of that understanding is formed in the context of evolutionary biology. Prior to evolutionary biology, a being such as a beaver might be classified in a natural order of things and might even be studied and observed. But today, due to the revolution of thought in biology that was crystallized by Darwin, asking "why" and "how" the beaver does what it does in the context of conventional biology necessarily involves the interrelated considerations of adaptation, genetics, and what we might call the present "story" of evolution that we rely on to make sense of the beaver's behavior. An example of this analysis is presented by Richard Dawkins and Yan Wong in *The Ancestor's Tale: A Pilgrimage to the Dawn of Evolution (2016),*

> Like any other useful adaptation, the dam-building clockwork in the brain must have evolved by Darwinian selection of genes. There must have been genetic variations in the wiring of the brain which affected dam-building. Those genetic variants that resulted in improved dams were more likely to survive in beaver gene pools. It is the same story as for all Darwinian adaptations. But which is the phenotype?[6] At

---

[6] According to Dawkins, "A 'phenotype' is that which is influenced by genes. That pretty much means everything about a body. But there is a subtlety of emphasis which flows from the word's etymology. *Phaino* is Greek for 'show', 'bring to light', 'make appear', 'exhibit', 'uncover', 'disclose', 'manifest'. The phenotype is the external and visible manifestation of the hidden genotype. The *Oxford English Dictionary* defines it as the 'sum total of the observable features of an individual, regarded as the consequence of the interaction of its genotype with its

which link in the chain of causal links shall we say the genetic difference exerts its effect? The answer, to repeat it, is all links where a difference is seen. In the wiring diagram of the brain? Yes, almost certainly. In the cellular chemistry that, in embryonic development leads to that wiring? Of course. But also *behavior*—the symphony of muscular contractions that is behavior—this too is a perfectly respectable phenotype. Differences in building behavior are without doubt manifestations of differences in genes. And, by the same token, the *consequences* of that behavior are also entirely allowable as phenotypes of genes. What consequences? Dams, of course. And lakes, for these are the consequences of dams. Differences between lakes are influenced by differences between dams, just as differences between dams are influenced by differences between behavior patterns, which in turn are consequences of differences between genes. We may say that the characteristics of a dam, or of a lake, are true phenotypic effects of genes, using exactly the logic we use to say that the characteristics of a tail are phenotypic effects of genes . . . The phenotype of a gene, in the true sense of the word, may extend outside the skin of the individual. Birds' nests are extended phenotypes. Their shape and size, their complicated funnels and tubes where these exist, all are Darwinian adaptations, and so must have evolved by the differential survival of alternative genes. Genes for building behavior? Yes. Genes for wiring up the brain so it is good at building nests of the right shape and size? Yes, by the same token, yes. Nests are made of grass or sticks or mud, not bird cells. But the point is irrelevant to the question of whether differences between nests are influenced by differences between genes. If they are, nests are proper phenotypes of genes. And nest differences surely must be influenced by gene differences, for how else could they have been improved by natural selection? Artefacts like nests and dams (and lakes) are easily understood examples of extended phenotypes. There are others where the logic is a little more . . . well, extended.

---

environment' but it precedes this definition with a subtler one: 'A type of organism distinguishable from others by observable features'".

In this passage, we see how the use of a mechanistic concept, "the gene", allows us to make sense of certain behaviors in way that we can relate to since we cannot question the source—i.e., the living being of the beaver—directly. Genetic reasoning appears to be the best we can do. Essentially, what we are doing in this mode of thought is linking up the external behaviors of the beavers (which we are treating as one object of thought) with the genes inside the beaver (which constitutes another object of thought) and establishing a new knowledge relationship, or order, which we then relate to and explore. This opens up new fields of inquiry where none previously existed. This knowledge relationship is established by us—not the natural world as given—through the direct observations of genes at one level and the inferences we draw from gross observations of the beavers at another. While the knowledge we have derived from this relationship must contain some undisclosed error (as all human knowledge does) it also contains some positive information as well and with this positive network of meaning we too can learn to build new things in the world, just like our friends the beavers do (although the "things" we are talking about here are in the nature of non-physical constructs, such as ideas).

Dawkins suggests that connections we instinctively draw between the characteristics, behaviors, and products of lower creatures and their genes are not so bright. This cautions us to not become over-reliant on our concept of the gene, lest we unwittingly succumb to delusion or error. When we look at a particular feature of the beaver, say its tail, we have little difficulty in saying that it is the product of a genetic function over which the beaver has little say or control. The expression of the dam, however, is a bit different because there is a sequence of behavior that the beaver must engage in over time before the dam becomes actual. Between the moment that there is "no dam" and the moment that the dam is created, there appears to be some sort of active, living self in the beaver which makes the dam happen. But is that dam purely an expression of a genetic function, or is there some non-genetically determined factor, such as "thought", that must occur to bring the dam into being? How would we make such a determination since we cannot "see" into the mind of the beaver itself and compare what is happening within the beaver, to what we understand is happening within us?

Dawkins' discussion of this case shows that drawing a bright line between the actions of the "living" organism, the genes that guide and direct the development and behavior of that organism, and impersonal forces of nature that cause things to happen is not always so clear. There is some controlling impulse in a modern human to tend to view the behaviors of the beaver as automatic and controlled by its genes—and, indeed, this impulse contains the seeds of a new form of knowledge, even if it also contains some undisclosed error or delusion. *See* ***Section 5.3.2*** (discussing how the delusional search for the "philosopher's stone" in Medieval Europe actually led to the development of real knowledge in the field of chemistry). But if one can say that dams are phenotypes of the genes of beavers, why can't one also say that computers are the phenotypes of human genes? Yet most humans would naturally assume that the computer represents a grand, superior achievement of human intellect and deliberation, not the product of genetic determination; and this is actually how the case seems to us from our perspective within the living selves of the human world. Is this simply because our thinking and own internal processes are familiar and known to us through our active selves, or is it because there is something objectively real about the things that we do which sets computers apart from dams in a meaningful way (aside from the fact that the computer is far more complex to construct)?

If we look at ourselves from an alien perspective, we might ask the following question: could it be the case, that, just as we see beaver behavior as being determined by a contributing force that is not the result of living, independent energy of the active self, that perhaps higher order aliens could see our activities in a similar way? It may be impossible for us to conceive of the precise way that a higher alien intelligence would see many of our behaviors as deterministic, just as the beaver could never understand the concept of the gene. But we can recognize that an intelligence which has evolved over larger cosmological time scales would likely operate at a scale of depth and time that is *far beyond* our simian imaginations, but *well within* unknown perspectives among the galactic order of intelligence. By definition, this type of intelligence would not make sense to us for the foreseeable future because we are just getting started as a scientific species capable of clearly apprehending such things. During the initial phase of our interactions with such beings, it is possible that they would see many things about the actions and limitations of human beings that they could not communicate to us any more than we could explain the workings of genetics to a beaver.

2.3 *Cosmic solemnity*

A realistic reaction we can have in such a scenario is to recognize the presence of a greater intelligence at work in our environment, while also recognizing that we simply cannot understand it. This would be a very new experience for our species, resulting in a dramatic change regarding all of our anthropocentric narratives. We could think of alien beings in this context as something akin to a galactic megafauna. In the terrestrial world, a human being can have the experience of being in the presence of a powerful being that could kill them, such as a shark, cobra, grizzly bear, or Komodo dragon. This would be analogous to this experience, but with an alien twist because these terrestrial creatures we have just mentioned are "lower" forms of life to us from the scientific perspective, whereas we would be the "lower" form of life in relation to galactic megafauna.

Such an attitude, if adopted, would caution us to have very low expectations regarding any direct transmission of knowledge from alien beings, and to restrain ourselves to the near-at-hand observations and experiences, instead of believing that we humans could be at their level, and would encourage us to adopt an attitude of cosmic solemnity towards the experience. *See* ***Section 6.6.1*** ("Why aliens should not be placed in the supernatural order"). Just as we cannot give a cogent lecture in genetics to a beaver no matter how much we may want to, it could be the case that some of these aliens simply cannot communicate certain things to us no matter how much they wish they could either. In such a case, the burden of increasing the possibilities of understanding and awareness lie with us, not them, and requires us to adopt a new attitude towards this experience and realization.

2.4 *Inquiry into the "inert" order of the invisible and created*

Returning now to the terrestrial level, when one considers the situation from a standpoint that does not assume that human consciousness is at the center of all forms of knowledge, drawing hard and fast lines on what sort of behavior is determined and what is not, can seem somewhat arbitrary. But our minds seem naturally drawn to these types of distinctions among causes involving living beings in an effort to create hierarchies of terrestrial orders among them. These hierarchies in turn create "places" for us in the real world which allow us to feel comfortable and settled. The example of the gene is one of the ways in which we do that, as it is sometimes used as a concept to differentiate our higher order activities from the lower order activities of creatures such as beavers when we make distinctions within the living order itself.

The inquiry into the "inert" order of natural objects is a bit different. We inquire into living and non-living orders from the point of view of the living self and place living things on one side of a divide, and "natural" objects on another. Inquiry into the world itself is alive for us because we are living, but yet this inquiry is also directed at natural objects that seem to originate independently of us, which do not immediately appear to have been created by a particular personality, and which we would not necessarily say are alive. In fact, we often times find "created appearing" objects which are initially hidden from our view and are only disclosed to us after much effort. These include things such as atoms, subatomic particles, viruses, germs, and even natural things that appear to be machines, such as the so-called "flagellar motor".[7]

We also discover laws which we use to explain various phenomena. When we discover things (such as atoms or molecules), we are struck with the sense that they have "been there all along", and yet we don't particularly see anyone there who has actually created them in the same way that we could find a particular human or an animal that might be responsible for a particular creation, such as a dam or computer. For instance, when we discovered the atom, it was almost as if a gift or surprise was discovered, yet we see "no one" around who we can say has actually left it. While we may rationally infer the existence of some type of Creator as a result of this kind of scientific experience, and attribute certain characteristics and powers to him/she/it/they, such an exercise is far different from the attribution of creations we make to other humans beings (or, in some rare cases, non-human terrestrial creatures).

---

[7] The flagellar motor is essentially a nanotech motor that exists naturally in a flagellum and moves its whip. This motor has a number of separate and discrete parts that are nearly identical to parts in motors that were much later designed by human beings who had no knowledge of the existence of the flagellar motor itself.

2.5 *The supernatural order*

Without getting drawn into a debate about theology, we can at least say that something seems missing in these situations between the world of our ordinary experience and what we seem to find when we examine nature in detail. It is almost as though we are seeing the aftereffects of the activity of some sort of presence which has now departed from the world. People sometimes react to this mysterious tension with an impulse to unite the "gap" between these orders so that things make sense as one whole. Sometimes this gap is closed by positing a third order, a supernatural realm beyond time and space, where special beings reside that have set the different orders in motion, govern those orders, and even sometimes directly intervene in them. These beings, which have been referred to as "God," "Allah," or "Yahweh" in some contexts, are not generally thought to be present in the world in the same way other beings such as people, animals, plants, and insects are. Instead, these supernatural beings have a more sporadic and irregular presence, appearing in the local realm from time to time and then suddenly disappearing. For many people, the existence of this invisible, supernatural order is indispensable and necessary for the world to make any sense at all to the point where they cannot even imagine a "real world" without the unseen, invisible one. Incidentally, this position is a rational one, based on the parameters of direct experience.

2.6 *The 3 orders*

Right or wrong, the relative separation of these three orders appears to be a fundamental feature of how we perceive the world in our era. In fact, it probably would be very difficult for modern rational humans to imagine a world in which there exists no firm distinction between living beings, natural objects, and supernatural entities. Although the approach of viewing living things, natural things, and supernatural things as forming separate orders is the dominant one in our era of thought, it is not the only way humans can see the world. In *The Copernican Revolution*, Kuhn notes that the early Egyptians, Babylonians, and Indians, articulated a cosmological framework that included the activities of humans and supernatural beings in a common order. For example, the sun was viewed by the Egyptians as "Ra", their principal god, and its motion across the sky was the actual behavior of this god that occurred in the world of human beings on earth, even though it was far removed. In this system, a third realm uniting human and natural orders was not needed because what we would see as three types of orders were all mixed into one. When the Egyptians saw the Sun, they were also directly seeing Ra, even though Ra was at a great distance from them. The reality of this kind of world is very difficult for the modern scientific mind to imagine and would constitute a non-ordinary type of experience (e.g., the type of experience a person might have while temporary under the influence of a psychedelic chemical agent, such as LSD).

Preliminary considerations of this nature are necessary in the present context because alien anomalies don't fit neatly into any of these categories, but appear to partake of certain aspects of both the human and supernatural, and are unlike the natural. But if we recognize that these categories are really just features of a paradigm, and not coextensive with reality itself, then we might be able to adjust them until we find a good fit for our alien anomalies. By contrast, if we solely rely on the parameters defined by the pre-existing categories themselves, we will likely remain stuck. Going back to the roots of these categories and bringing to awareness the underlying thought processes which bring them into being can help us see how they are not as firm as we think they are.[8] But if we fail to examine conventional categories and we always accept them as givens, as something which we have no choice but to use, we will have great difficulty seeing beyond them and seeing how they can be changed, just like Copernicus' detractors could not see a real world in which the earth actually moved in space.

These basic patterns discussed above are the dominant approaches that we use and will rely upon when we try to think about and understand aliens. We can't simply change conventional categories with the snap of a finger, they have weight, momentum, and gravity within the field of thought and perception itself. But if we believe too truly in the categorical limits created by conventional thought itself, we will be misled time and time again, because alien presence, behaviors, and effects blurs the lines between things that are living, supernatural, and unreal. These

---

[8] Recall how, in the Copernican Revision, those who believed the earth was at the center of the universe stubbornly clung to their false paradigms no matter what, believing that they were real. They did not realize that the Aristotelian paradigm was just another paradigm, and that all paradigms can be changed.

anomalies simply "don't fit" at all, which then causes the unreflective mind to immediately reject or suppress them, or engage in unwarranted fantasy speculation about them. A different approach, however, is to question the primacy and superiority of the categories themselves, or to at least become aware of them *as human categories of thought* and not simply features of the world "as it is", with the prospect that they might be adjusted or altered in a certain way.

2.7 *The supernatural perception of aliens*

Due to the characteristics of the structures of the conventional categories themselves, and how they have been applied by humans to the world for generations, when humans approach the idea of aliens, let alone what an individual human might consider to be an individual experience of an alien, they are perceived primarily as supernatural beings. While this is the prime category that may initially have the "best fit," it actually does not fit at all because an actual alien—one that really evolved on another planet and made its way to earth—could not, by definition, be a "supernatural being" as conventionally conceived by humans. The perception of aliens as supernatural beings necessarily distorts our ability to think about them clearly due to our prior associations and ideas about supernatural beings, ideas which were formed in the infancy of our species and have a poor chance at assisting us in this new realm of understanding.

We understand that aliens certainly would display features that appear to the human mind as being "god-like", supernatural, divine, or magical. Because of this, the mind will initially place them in this conventional category as a first impression and then continue to think of them in that way. But when rationally considered, it is self-evident that aliens simply cannot be "gods" or "supernatural" beings in the manner in which those categories have been conventionally developed and deployed within our species. The categories of thought we rely upon are obviously built up over time through a process of individual and collective learning and experience. But the sources giving rise to these paradigms contain no real knowledge of aliens that would be appropriate to this era of human thought since aliens have never been studied as scientific objects.[9]

---

[9] For a more detailed discussion of the alien experiences issue, please see **Section 6.0**.

Treating aliens as gods or supernatural beings may make sense as a limited first impression (especially thousands of years ago), but it would be error to continue to think of them in that manner in today's time. Today, instead of thinking of aliens in supernatural terms, it is more accurate and realistic to begin thinking of them in from the standpoint of scientific disciplines such as evolutionary biology. This is the rational starting point implied by conventional categories of thought which clearly admit of the rational possibility that intelligent aliens likely evolved on other planets in the universe.

If we understand the lessons from the Copernican experience, we see how at this point in time, it does not make sense to have too many ideas about aliens, just as it would have been foolish for Copernicus to attempt to understand gravity, acceleration, velocity, and a whole range of other alien paradigms about the physics of the world *before* he could feel comfortable about the Copernican Paradigm itself. *See* ***Sections 4.3 and 4.4*** (discussing the pendulum anomaly, paradigm shifts in clinical medicine and explaining how excessive ideation can interrupt the development of real knowledge). If we place our faith in the dynamics of the Copernican Revision, then we must conclude that ideas about aliens will remain in a "negative space" in the field of thought at the start, since we have no data, information, or knowledge from a real study about them. In other words, if we are to realistically think about aliens, we should do so from the standpoint of simply seeking to outline the realistic possibilities for their appearance at first, rather than attempting to determine their concrete characteristics (which would best be determined by measurement or real scientific observation at a later date).

In practical terms, this inclines us to recognize that if our minds impulsively treat aliens like gods or supernaturally wise beings, it may be due to an unreflective and erroneous application of conventional categories themselves, a distortion of things as they are. We know that human species has a long history of projecting and speculating as to what such supernatural beings might be like in the absence of direct evidence. But such speculation runs a high risk of distorting real knowledge about aliens as developed through a scientific process (this is a point which will be demonstrated by reference to the development of modern clinical medicine at a later point in this work).

A significant aspect of the enterprise of gaining real knowledge in a situation like this (i.e., where a persistent anomaly fails to fit in a conventional category of thought) is to understand what one *shouldn't be doing* just as much as what one should. This is because the ongoing delusional application of conventional categories will create a self-defeating doom loop whose unfolding in space can only be arrested through non-action (instead of obsessively trying to make a square peg fit into a round hole by continuing to perpetuate these thoughts). In this connection, we know that collective thought patterns are badly misaligned with respect to the presentation of the alien anomalies themselves, and have not corrected themselves in a way that resonates productively with those anomalies, because there has been no real change in the field of knowledge in over 60 years. *See **Section 2.10*** ("Emergence of alien anomalies as non-visionary phenomena"). Part of the reason for this situation is because we tend to treat aliens like supernatural beings instead of genuine extraterrestrials who have actually evolved on other planets and who could be studied in the same type of way that we study other types of creatures we encounter through a scientific process.

2.8 *The 3 orders and scientific thought*

The mode of thought whereby separate orders are related to one another through the activity of a living self can be traced back to the Greek philosopher Anaximander of Miletus, sixth century B.C., which may be the earliest recorded example of what we might call "scientific thinking".[10]

> The stars are compressed portions of air in the shape of [rotating] wheels filled with fire, and they emit flames at some point from small openings . . . The sun is a circle twenty-eight times the size of the earth; it is like a chariot-wheel, the rim of which is hollow and full of fire, and lets the fire shine out at a certain point in it through an opening like the nozzle of a pair of bellows . . . The eclipses of the sun occur through the orifice by which the fire finds vent being shut up . . . The moon is a

---

[10] There is no real reason to think that cultures outside of Ancient Greece did not also develop methodologies of a similar nature. But whether due to the laws of chance, or other historical accidents, our present scientific mindset does actually trace its roots back to this particular lineage of thought.

> circle nineteen times as large as the earth; it is like a chariot wheel, the rim of which is hollow and full of fire, like the circle of the sun, and it is placed obliquely, as that of the sun also is; it has one vent like the nozzle of a pair of bellows; its eclipses depend on the turning of the wheel.

Obviously, Anaximander's factual conclusions in this fragment appear "incorrect" to the modern scientific mind, but the method of analysis and the manner in which causes are separated still has the ring of truth and resonates with our active paradigms. In this style of analysis, information available to the senses is examined in a sequential format, with attempts to analogize among different orders of perception. Anaximander's description attempts to relate things which are poorly understood (the sun and moon) to things that are familiar (such as chariots and pairs of bellows), and in this interchange of analogy new orders of knowledge arise.

While Anaximander's description may sound childlike or fantastic, it is not entirely wrong either. Clearly, Anaximander's direct analogies between chariots and bellows is a misleading description for what we today would consider the actual mechanism at play here. Furthermore, we today certainly take issue with the results of Anaximander's measurements of the relative sizes of the celestial bodies. At the same time, Anaximander's observation that the stars are "compressed portions of air" is not entirely inaccurate. While he may not have had the details of this compression correctly stated due to the limits inherent in his mode of observation and the ideas that existed in his time, he was certainly on the right track to recognize that the stars had formed as the result of a process of compression involving other elements drawn from the space around them. This seed of thought ultimately ripened into real knowledge over a long period of time.

A more modern of example of this approach is described by Bohm in relation to the insight into the universal nature of gravity. As Bohm relates, Newton's initial insight into gravity came with an imaginative intuition that the moon was falling, even though it never quite reached the earth. At first, this flash of insight was only implicit, not clearly worked out in all of its details, taking the form of a "mental image which, as it were contains the main features of the new perception spread out before our 'mental vision'". But as the insight was contemplated, Newton tried to make the formulation more precise in order to figure out, for instance, just how fast the moon was falling. Bohm postulates a possible way in which this subsequent, more precise formulation, was derived in the mind of Newton.

> [F]rom the data known to him, Newton was able to show that the moon falls considerably more slowly than an object on the surface of the earth. This meant that the force of gravitation must decrease in some way with the distance. But the question was: "Just how fast does it decrease?" It is quite possible that the precise form of the hypothesis adopted by Newton may have been suggested by calling to mind some already available image, such as that of the light intensity from a radiant object, which was already known to fall off as the inverse square of the distance. So then, the thought would arise naturally: "As with light, so perhaps with gravitation". Even if in actual fact it may not have been this image, but perhaps some other one, that suggested the inverse square law, the essential point remains that hypotheses generally involve new forms, arrangements, connections, and meanings of images already available in the mind. So, a hypothesis is primarily a form of fancy or, of constructive thought, the validity of which has to be continually tested by further appeals to observed fact.

Knowledge developed in this way is articulated into ideas and theories which form a narrative of observation, or what is sometimes referred to as a "paradigm". This narrative itself then becomes an object of study and is subject to constant revision, until it eventually becomes detailed enough to be measured. *See* ***Section 2.17, Footnote 18*** (discussing the role that ever-increasingly precise measurement plays in the development of scientific knowledge). Some narratives stand the test of time (such as the universal nature of gravity, which still remains valid under certain sets of limiting conditions to this day, or the plays of William Shakespeare which still describe the paradigms of human drama in ways which the audiences of today can directly relate to), while others fall (such as Anaximander's hypotheses regarding celestial phenomena, which sounds like fantasy to us now, but clearly made sense to Anaximander when articulated). In this approach, the mind is naturally drawn to the phenomena of the world and seeks to accurately comprehend the causes of as many classes of phenomena as possible. Although the content that we work with now is far different than it would have been in the time of

Anaximander, this underlying style of analysis remains fundamental to our outlook on the world to this day.

2.9 *The real, the unreal, and the Copernican Trap*

In addition to the discrimination between the three orders and living/non-living distinctions, we also make discriminations between things which are real and things which are not. For instance, science fiction movies sometimes portray humans and aliens interacting, which is something that doesn't seem real to the vast majority of people. Let us assume a Scenario A where there is no doubt that the movie itself is real but that the portrayed interaction is not. This is a case where it is easy to separate the real from the unreal. It arises from the context under which the movie is viewed. The viewer pays money to gain entrance to a movie theatre, and then watches a spectacle unfold which they know is not real. The person going to the movies understands that they saw "fake aliens" in a movie that used real actors and cost real money to make. The viewer does not think that the content she is watching when she is in the theatre is "actually real".

However, let us assume a Scenario B where someone has unknowingly been given LSD and starts tripping while watching such a movie and believes the interaction is real. Or assume, perhaps, that the movie depicting a fictional alien-human encounter is staged in such a way to make the viewers believe it is real without knowing it is not (a variation on Scenario B actually did happen with Orson Wells' radio presentation entitled *War of the Worlds* (1938)). In these cases where the fictional scenario is mistakenly believed to be real, the objective features in the presentation do not change, but the structure of the moment of between viewer and movie does. We could say that the subsequent experience at Time 1, is conditioned by the structure of that moment in advance at Time 0. We can then study the conditions of formation at Time 0 to learn more about how those expectations and conditions effect future events for a human. In Scenario A, the viewer knows in advance that the movie is not real and so has a certain class of experience in the moment of viewing. This moment might be described as having the character of entertainment, enjoyment, or education. Perhaps the movie makes the viewer think about things in a new way, but it does not change the perceived understanding of the reality of her world. The case is different in Scenario B, when the viewer delusionally thinks the movie is reality for one reason or another.

The determination as to whether an oncoming experience is real or not will have a tangible and concrete effect on human reality itself. Assume that prior to the moment of viewing, the viewer in Scenario B had conventional beliefs about alien and humans having contact. Afterwards, those beliefs are changed in an abrupt shift when the viewer in Scenario B thinks that they just saw a "real" interaction between a human and an alien. The shift could take many different forms: wonder, amazement, terror, panic, obsession, nonchalance, etc. The future effects from Scenario A or B are different, however, regardless of how or why they are different. In Scenario A, the current relationships of the world remain largely undisturbed at that moment, but perhaps a new way of thinking about things has emerged. By contrast, Scenario B has an immediately disruptive effect, upsetting the conventional perceptions of the world in the moment itself. *See* ***Section 6.4*** (discussing the Pascagoula Encounter). The difference in effects is clearly attributable to the effects of different judgments within the human self. While we may not know exactly what is going on in these types of moments, we do know that these types of changes are quite real.[11]

This basic thought experiment demonstrates that the manner in which a judgment is derived can determine the power and the effect it will have on the individual mind and subsequent attitude towards the world. Because of our technological advances, we now are able to create scenarios and experiences which may seem entirely real but are not. This lived experience for modern humans gives us different ways of understanding how knowledge is created and how the ideas of what is real can be played with in different ways. Furthermore, it has been shown quite conclusively throughout recent history that when we induce a sense in ourselves that a certain scenario is real—even though it is not—we can create a real action which can create its own real effects even though it may be built on the scaffolding of delusion. Examples of this include the famous Stanford Prison Experiment (1971), Hitler's Third Reich (1933-1945), and the creations of Derren Brown in *The Push* (2016), *The Sacrifice* (2018), and his other works as well.

---

[11] The discipline of neuroscience can delve into the minutiae of these situations and conditions in greater detail.

These particular sets of considerations are important to bear in mind because the possibility of real aliens having contact with humans in the present time is the type of event that conventional understanding instinctively categorizes as "unreal" for the reasons cited above. This feeling of unrealness then instinctively inclines the human mind to reject the occurrence as a real possibility before a form of scientific examination has had an opportunity to conduct an inquiry. This is the same trap that the overwhelming majority of humans fell into during the Copernican Times. Today, we look back upon their instinctive rejection of these ideas and may be critical of our ancestors for their lack of open mindedness—but is the same thing possibly happening today within the scientific community on the issue of present-day human to alien contact?

When the people in the era of Copernicus did not believe that the earth moved around the sun, their relationship to the ideas of Copernicus were likely similar to how one might relate to a work in Scenario A, but with a critical twist; insofar as they falsely convinced themselves that Copernicus' ideas were pure fiction, and related to them as such, they in fact turned out to be real. Put differently, we might say that they were watching a documentary but told themselves it was fiction in order to preserve other aspects of the conventional narrative which held their real worlds together. From our point of view, their logic seems twisted; to them, it made total sense.

The conventional reaction of people today to the people of Copernicus' Time, is to say that we are "smart" and they were "dumb", thus assuring ourselves that we would have recognized right away the truth had we ourselves been alive in those times. But this is a simplistic and unsophisticated way to view the situation. Copernicus' ideas conflicted with the facts of lived experience themselves. Those facts seemed objectively real and provided a basis upon which the humans of that time ordered their entire existence; they had a powerful inertial effect on human thought and perception, even though they were shaped by a false, Aristotelian paradigm. *See* ***Section 4.3*** ("the pendulum anomaly"). This paradigm had a force and inertia which was not easily overcome, and most fell victim to its power. A similar thing is possibly happening today with respect to the issue of alien presence today. The feeling of unreality that this idea carries with it is simply too much for most people. *See* ***Section 6.1*** ("the intrinsic strangeness of alien experiences"). Most people today, just like Copernicus' detractors, get carried away by this feeling and fail to engage with it in a scientific process. *See* **Section 5.3.4** ("Avoiding anti-Copernican errors"). This, at least, is one of the core hypotheses animating this text.

2.10 *Emergence of alien anomalies as a "visionary phenomena"*

One of the powers of the scientific approach is its ability to take things from the world of the invisible and absurd and bring them into conventional systems. The modern world is filled with objects that the powers of science have moved from the category of the unreal, into that of the real, and then placed within an arrayed, conventional order, that we get to work with directly instead of only inferentially. These include things such as germs, viruses, atoms, DNA, black holes, and an endless list of other objects. We can have faith in the lineage of knowledge that has developed in this way, believing that if we apply it to alien anomalies, they could also one day be part of the real world of humans, just as was the case with the Copernican Revision's incorporation of the true motion of the planets into the human vision of outer space.

As suggested in the Introduction, throughout human history there have existed a genus of anomalous phenomena which defy easy classification into either natural or human causes and cannot be placed easily within the bounds of a real phenomenon. In prior eras, these anomalies would necessarily be placed in the supernatural order (or considered to be "not real") as humans lacked any other classification with which to relate to them. The etymology of the term "anomalous" traces its roots to the Greek word "anomalos", which translates as something that is uneven or irregular, deviating from what is normal. The term "normal" is itself derived from the Greek word "nomos" which generally refers to the "body of law" governing a certain form of behavior or action. The judgment of whether something is irregular is always formed by reference to what is thought to be normal in advance, based on prior experience or thought, or what we think should be appearing. An anomaly is, in a sense, something that should not be appearing, or which was unexpected, unexplained. The judgment of what was to be expected is necessarily formed before the anomaly arises by reference to ideas, concepts, or expectations that were embraced by the observer prior to the moment when the anomaly was presented. Following the work of Kuhn, an anomaly can also be defined as a phenomenon which “fails to perform in [an] anticipated manner” and “despite repeated effort” cannot “be aligned with professional expectations.”

An ancient example of this kind of anomaly is described in the Book of Ezekiel. This book describes a scene where a shiny cloud carrying certain “supernatural” creatures appeared to Ezekiel as follows:

- *Also out of the midst thereof came the likeness of four living creatures. And this was their appearance; they had the likeness of a man.* Verse 5.
- *As for the likeness of their faces, they four had the face of a man, and the face of a lion, on the right side: and they four had the face of an ox on the left side; they four also had the face of an eagle.* Verse 10.
- *The appearance of the wheels and their workings was like unto the color of a beryl, and all four had the same likeness; and their work was, as it were, a wheel in the middle of a wheel.* Verse 16.
- *And when the living creatures went, the wheels went by them; and when the living creatures were lifted up from the earth, the wheels rose.* Verse 19.
- *And above the firmament that was over their heads was the likeness of a throne, as the appearance of a sapphire stone: and upon the likeness of the throne was the likeness as the appearance of a man above upon it.* Verse 26.
- *As the appearance of the bow that is in the cloud in the day of rain, so was the appearance of the brightness round about. This was the appearance of the likeness of the glory of the Lord. And when I saw it, I fell upon my face, and I heard a voice of one that spake.* Verse 28

For Ezekiel personally, he interpreted what he saw as a manifestation of the supernatural being Yahweh. For some modern observers, it is believed that this may have been a report of an actual alien spacecraft, interpreted through the eyes and mind of someone with no grasp of modern science. Was it the product of an actual external occurrence, the result of delusion, hallucination, or cognitive error, or something in between? Regardless of what Ezekiel saw, the description is anomalous, out of the ordinary, irregular, difficult for the human mind to easily explain. This is not the kind of thing a human being sees as a normal part of terrestrial regularity.

The next most notable examples of this type of anomaly are reflected in the Nuremberg and Basel Broadsheets which were brought to the world's attention by C.G. Jung in his monogram, *Flying Saucers: A Modern Myth of Things Seen in the Sky* (1958). On April 14, 1561, the residents of the cosmopolitan city of Nuremburg were treated to a "very frightful spectacle" at sunrise involving "'globes' of a blood-red bluish, or black colors, or 'plates' in large numbers near the sun, 'some three in a row, now then four in a square, also some standing alone. And amongst these globes some blood-coloured crosses were seen'." This event was memorialized in a document now known as the "Nuremberg Broadsheet," which is available for viewing on the internet. A similar thing was observed over Basel, Switzerland on August 7, 1566. On that date, the citizens of Basel reportedly witnessed "many large black globes" in the sky, "moving before the sun with great speed, and turning against each other as if fighting. Some of them became red and fiery and afterwards faded and went out." This too was memorialized in a woodcut called the "Basel Broadsheet" which is also easily accessible on the internet.

The Nuremberg and Basel sightings occurred during a time when human perception and cognition was undergoing a critical change in trajectory which ultimately resulted in the mindset that we today consider scientific. It was during this time period, the mid-16th century, that humans began to analyze themselves, and the world around them, in a manner that looked for the deepest patterns of knowledge and reflection in an analytical fashion. This body of knowledge ultimately transformed into Newtonian physics, Einsteinian physics, clinical medicine, evolutionary biology, and the world of quantum physics and the computer in which we find ourselves today. If we wanted to, we could show how modern scientific thought was born during this time. One could say that the Nuremberg and Basil sightings form the earliest recorded example of this class of anomaly within the range of modern scientific thought.

From the mid-16th century to the mid-20th century, scientific products exploded. With the advent of World War II, humans began spending more time in the sky as the powers of aviation technologies were pressed into the service of modern warfare. During this time, pilots of the era reported seeing what they later termed "Foo Fighters" during their aeronautical travels. According to legend, the term was first coined by the 415th Night Fighter Squadron in November of 1944 to describe strange and mysterious aerial phenomena that they noticed during their flights. At the time, the term "foo" was a colloquial term used to indicate something that was nonsensical.

After World War II concluded, the use of aviation technologies continued to expand. In 1947, an amateur pilot from Idaho named Kenneth Arnold was flying a CallAir A-2 airplane over Mineral, Washington, and claimed to see airborne objects flying in "a diagonally stepped-down, echelon formation," stretching out over a distance of roughly five miles, on a single, horizontal plane, weaving from side to side, flipping and banking and darting around, like "the tail of a Chinese kite." Arnold later described their movements to reporters "like a saucer if you skip it across the water"; thus, the term "flying saucer" entered the lexicon of the popular imagination. Eventually, these objects were called unidentified flying objects, or "UFOs". Today these types of anomalies are sometimes called "UAPs", or unidentified aerospace phenomena.

Apparently, Kenneth Arnold was not the only pilot coming into contact with these types of anomalies in their aeronautic travels. In 1947 the U.S. Air Force initiated "Project Sign", a classified program whose purpose was stated in the Introduction to the technical report (which was officially declassified in 1961 and is widely available on the internet) as follows,

> This report is written to present the status of work being accomplished on Project "Sign", to summarize the data collected on sightings of unidentified aerial objects, to review the methods and reasoning applied in the evaluation of the data, and to present the results so far obtained from the study of data available.
>
> It is not expected that this report can present a final estimate of the situation regarding all the incidents reported. The data is still being studied by specialists in the fields of astrophysics and psychology, and further information is being collected to enable personnel evaluating project "Sign" incidents to

determine the possible explanations of some of the sightings. However, the report will furnish information on the present state of the investigation to staff personnel in this headquarters and in high echelons, and to others who are required to assess the possibility of a threat to national security presented by the sighting of such large numbers of unidentified flying objects.

A pursual of the Project Sign report demonstrates that the Air Force viewed these anomalies as real objects that were unwelcomed and interfered with their work. The circumstances attendant to the report itself provides convincing *prima facie* evidence of a real world reaction to a real world occurrence of some sort. The report examines a total of 243 incidents and groups the anomalies into four basic classifications: flying discs, torpedo/cigar shaped bodies with no wings or fins, spherical or balloon shaped objects, and balls of light. Project Sign involved a wide range of experts to examine these anomalies, including astro-physicists, meteorologists, engineers, psychologists, and even the FBI. When an anomaly could be explained by reference to known causes, such as vertigo, "optical illusions," or "errors of the human mind and senses", it was. Yet, the report does not assume that because some anomalies had a prosaic explanation, all of them necessarily did. The Project Sign report did not come to any firm conclusions with respect to the ultimate causes of these anomalies but was more akin to a preliminary inquiry focused on trying to accurately describe what was actually being seen in the skies. "No definitive and conclusive evidence is yet available," the authors write, "that would prove or disprove the existence of these unidentified objects as real aircraft of unknown and unconventional configuration. It is unlikely that positive proof of their existence will be obtained without examination of the remains of crashed objects. Proof of non-existence is equally impossible to obtain unless a reasonable and convincing explanation is determined for each incident".

Clearly, the examination of this subject was considered a matter of serious importance by those who were charged with protecting the skies over America at the end of World War II and the dawn of the Cold War. This, combined with the fact that the report was initially classified, is extremely powerful evidence that the U.S. Air Force truly believed that something real was going on which they did not understand, and which they needed to understand as a matter of operational security. No rational observer of the past could conclude that the Air Force was being afflicted by mass delusion in this regard. Something real was going on, something was being seen that was not understood and that was interfering with the Air Force's operational ability in a real way. This was not something that the Air Force believed it could simply ignore. In this respect, the report contains the following recommendations,

> Future activity on this project should be carried on at the minimum level necessary to record, summarize and evaluate the data received on future reports and to complete the specialized investigations now in progress. When and if a sufficient number of incidents are solved to indicate that these sightings do not represent a threat to the security of the nation, the assignment of special project status to the activity could be terminated. Future investigations of reports would then be handled on a routine basis like any other intelligence work.
>
> Reporting agencies should be impressed with the necessity for getting more factual evidence on sightings, such as photographs, physical evidence, radar sightings, and data on size and shape. Personnel sighting such objects should engage the assistance of others, when possible, to get more definite data. For example, military pilots should notify neighboring bases by radio of the presence and direction of flight of an unidentified object so that other observers, in flight or on the ground, could assist in its identification.

For reasons that are unclear, the Project Sign approach was never followed. Instead, there was a counterreaction by the U.S. Air Force which suppressed these anomalies and denied they existed. Although one could cite many instances of this reaction, a paradigmatic example is found in the case of Nick Mariana's "Montana Movie," which is documented in Joan Bird's work, *Montana UFOs and Extraterrestrials* (2013).

In 1950, just one year after Project Sign officially ended, Great Falls, Montana, resident Nick Marianas was somewhat of a Montana celebrity. Marianas was originally from Miles City, Montana, and was allegedly offered a contract to become a pitcher for the Brooklyn Dodgers out of high school. Marianas declined the offer to play baseball, pursued a college education and graduated from the University of Montana School of Journalism. Marianas was a well-known sportscaster in Montana with his own television program, radio broadcast and newspaper column, as well as the general manager of the Great Falls farm team "The Selectrics". By all accounts, Marianas was a well-respected member of the Montana community.

On August 15, 1950, Marianas was at the clubhouse at Legion Field in Great Falls, when he happened to notice what he later described as follows,

> I glanced northward to the tall Anaconda Company smokestack to check the direction of the wind from the white smoke. Force of habit, I suppose because our outfielders use it as an indicator on defensive play. As I looked up, I saw two silvery objects moving swiftly out of the northwest. They appeared to be moving directly south. The objects were very bright and about 10,000 feet in the air. They appeared to be of a bright shiny metal, like polished silver. Both were the same size and were travelling at the same rate of speed, which was much slower [than] the jets which shot by shortly after I filmed the discs. Suddenly they stopped. It was then I remembered the camera in the glove compartment of my car. I raced downstairs, yelling for my secretary, Miss Virginia Raunig. The distance from the top of the stairway to my car is about 60 feet, and I must have made that in about six jumps. I asked my secretary if she saw anything, and she said, "Yes, two silvery spheres." I unlocked the glove compartment of my car, took out the camera, turned the telephoto lens on the turret into position, set the camera at F-22, picked up the objects in the viewfinder, and pressed the trigger.
>
> The discs appeared to be spinning like a top, and were about 50 feet across and about 50 yards apart. I could not see any exhaust, wings, or any kind of fuselage. There was no cabin, no odor, no sound, except I thought I heard a whooshing sound when I first saw them. As the film clicked through the camera, I could see the objects moving southeast behind the General Mills grain building, and the black water tank directly south of the ballpark. I filmed the objects until they disappeared into the blue sky, behind the water tank.

The "Marianas' movie" may be the first publicly known film of one of these anomalies in human history. At the prompting of a friend, Marianas sent the film to the Commanding General of the Air Force at Wright-Patterson Air Force base in Dayton, Ohio, the same place where Project Sign was initiated. The Air Force accepted the film and then removed portions of the film which reportedly showed the aerospace anomalies before returning it to Marianas. When Marianas questioned the missing segments of the film, the Air Force denied that it had altered the film. Marianas was greatly distressed by the response of the Air Force, which caused a controversy that was later the subject of a congressional investigation. Before Marianas sent the film to the Air Force, however, he had shown it around Great Falls and it was reported in the local media. Apparently, the film was quite impactful on those who saw it. In her book, Bird quotes Marianas who said that after he showed the film to about 175 people at the local Lion's Club, "you could hear a pin drop".

Setting aside the exact nature of what it was that Marianas saw, we can see a vast difference in the type of description that Marianas presents when compared to that of Ezekial or the two Broadsheets. The Marianas description is precise and physically detailed. It provides a sense of perspective and measurement, gauging the presentations of the anomalous objects by reference to familiar and stable landmarks and other objects. The Marianas description paints a picture which is clear, precise and relatable to the modern mind in a scientific way. We see no reference to supernatural beings here. While Marianas was highly excited, he was not overwhelmed and was able to record and measure. We see an evolution of our perception of these anomalies in these sparse examples.

Another thing we notice in this example is the suppressive response to the anomaly which continues to serve as a persistent barrier to the development of our knowledge today. This is a real problem because the U.S. military has the best evidence of alien anomalies due to the types of activities it conducts and where it conducts them.

The Air Force's response to the Marianas film was the opposite of the open collaborative approach suggested by Project Sign. In fact, and as is well known, from the time of Project Sign and the Marianas' film until the start of the 21st century, knowledge of these aerospace anomalies was actively suppressed by the U.S. military and other government agencies in America. Instead of being treated as exotic objects one could study (such as atoms, germs, or other similar phenomena), they were treated as non-objects or rumored objects. In his 1958 monograph *Flying Saucers: A Modern Myth of Things Seen in the Skies*, C.G. Jung referred to these as "visionary rumors". A visionary rumor differs from an ordinary rumor in that an ordinary rumor is driven by "popular curiosity and sensation-mongering," whereas a visionary rumor hits a deeper connection within humans and stirs strong and compelling emotions. Jung cites historical examples of such visionary rumors to include collective visions witnessed by the Crusaders during the siege of Jerusalem, the troops at Mons in the First World War and various other religious visions. Unlike religious visionary rumors, the visionary rumors circulating around these modern-day aerospace anomalies display a technological and scientific bent as opposed to a religious or spiritual one. Jung says these anomalies appear to be a "physicist's miracle", since they seem to be real technological objects which can do things that defy all the normal laws of physics and reality as we understand them.

We can safely say that from the post-World War II era until the present, these alien anomalies have been treated far differently than other exotic objects that humans might study. There seems to be something about the mere possibility of an alien being present in the physical space of our planet that causes many people to simply lose their minds, stirring strong and powerful subconscious reactions which are not easily overcome; although for others, these objects are a source of great curiosity, delight and interest. It is not clear why those who are genuinely interested fall into the minority camp. In any event, because only a minority of humans are interested, and not fearful, of these anomalies, the conventional collective reaction has been to suppress or deny them instead of actively investigating them with a strong and determined sense of curiosity.

The status of these anomalies as visionary rumors has meant that they have not been handled as serious scientific facts, although they did become a part of popular culture, art, myth, and legend. We can see that this is the case when we consider the knowledge of these anomalies that Jung described in 1958 and compare that knowledge to what we know today. A summary of the key points taken directly from Jung's description in 1958 is as follows:

- There is a high public interest in the anomalies.
- They're usually round, metal, and mechanical.

- They can't all be explained away as birds, balloons, balls of fire, clouds, products of delusion or human error.
- They range in sizes.
- They seem sometimes to fly like insects.
- They appear to defy gravity.
- They sometimes show up on radar, while not visible to the naked eye.
- Sometimes they are visible to the naked eye, but do not show up on radar.
- No humans could survive the rates of speed and motion displayed by these objects.
- Their movements cannot be explained by reference to our current understanding of physics.
- They're interested in airfields, industrial installations, and nuclear missile sites for some reason.
- They sometimes seem like groups of tourists "unsystematically viewing the countryside".
- Their movements appear purposeful and intentional to us, giving rise to descriptions such as "evasive" or "defensive".
- Sometimes they are reported to have occupants.
- They seem to engage with the military.
- There are stories of direct encounters between them and humans.
- There is gross speculation on the reasons why they are here (e.g., to save humans from nuclear war or other forms of technologically driven catastrophe).

The foregoing summary of knowledge from Jung's work in 1958 is a very close fit to what we know about these anomalies today.[12] As you can see from reading this description, it is confusing, contradictory, and just doesn't make sense. This is a strong signal that this field of study has not yet ripened into conventional knowledge (conventional knowledge is, by definition, knowledge that feels settled and "makes sense"). As Jung himself complained in 1958,

> Unfortunately, after more than ten years' study of the problem I have not managed to collect a sufficient number of

[12] Some would disagree with this statement and argue that there is a wealth of knowledge about these anomalies that is held in secret by various military agencies.

observations from which more reliable conclusions could be drawn. I must therefore content myself with having sketched out a few lines for further research. Of course, next to nothing has been gained as regards a physical explanation of the phenomenon.

2.11 *Emergence of alien anomalies as "non-visionary" phenomena*

From the time that Jung wrote these words until the end of the 20th century, alien anomalies remained unreal objects from the standpoint of conventional human experience. While all sorts of rumor, mystery, and intrigue surrounded these phenomena, they were certainly not considered normal objects from the standpoint of conventional thought and experience. They remained exiled to the realm of visionary existence. As a result, no real knowledge regarding these objects was developed. We could analogize this situation to that of the atom which was completely speculative and fanciful from the days of Lucretis until it was actually discovered by Rutherford using modern scientific instruments. From that point forward, however, how we handled the atom was far different and our knowledge about it changed as well.

Those of us living today are witnessing the emergence of a similar situation with respect to alien anomalies. Through a gradual and cumulative process that gives the fearful simian mind a chance to adjust to new conditions of life, modern methods of observation are now allowing us to see alien anomalies as real and normal objects instead of ones that are simply visionary or imaginal.

We can fix the beginning of a new era of collective understanding with the publication an article entitled *Glowing Auras and 'Black Money': The Pentagon's Mysterious U.F.O. Program* in the New York Times in December of 2017. The article confirmed that the U.S. government had been funding research into unidentified aerial anomalies, including programs examining physical materials that had been recovered. United States Senator Harry Reid, who was a driving force behind the program, had even informed high ranking members of the U.S. military that the government programs had made "much progress" with the "identification of several highly sensitive, unconventional aerospace-related findings". The article was also accompanied by the release of video that was confirmed as authentic by the U.S. military, showing some of the anomalous occurrences which were under study.

Glowing Auras demonstrated that human technological capabilities had advanced to the point where these anomalies could be observed over a time and space that had a chance of yielding real knowledge, instead of the strictly visionary forms that had dominated before. As we noted in the discussion of the beaver, the ability to observe beaver behaviors over an extended period of time, under a variety of different conditions, is the necessary precedent to giving us the information we need to draw meaningful deductions about the causes of those behaviors and internal processes. The *Glowing Auras* article showed us that the same is now true with respect to alien anomalies. Regardless of what these alien anomalies really are, *Glowing Auras* showed that it was possible to generate real knowledge and information about their general characteristics (which, of course, would constitute a first step of study in any science before greater knowledge and details are established). Information at this level of detail and precision was simply not possible in prior eras.

Unfortunately, this data and information was not brought into the world of general knowledge and was suppressed, inhibiting the development of understanding. As we saw in the case of the Marianas Movie, this course of suppression has been in motion for quite some time. Nevertheless, *Glowing Auras* made it clear that alien anomalies were real occurrences that could actually be studied by modern scientific means. Again, this is similar to what happened when the existence of the atom was actually confirmed by Rutherford; it didn't mean that all the details were known, but simply that it was confirmed as legitimate object of study going forward. This confirmation then provided the scientific conviction necessary to invest the time and resources necessary in order to learn more.

2.11.1 *The tic-tac anomaly*

In the wake of *Glowing Auras* a few documented encounters between United States service members and anomalous objects were publicly revealed and confirmed by the United States government as being real. The most substantive and well documented of these encounters occurred on November 14, 2004, between service members of the U.S. Navy's Carrier Strike Group

Eleven (CSG 11) and an object approximately 40 to 50 feet in length, 10 to 15 feet wide, white in appearance, resembling a giant "tic tac," a popular candy item in America. Evidence of this occurrence comes from three primary sources: 1) eye-witness accounts of multiple service members; 2) circumstantial evidence from logbooks and other government documents; and 3) direct evidence of the object as captured on a 76-second video taken on an AN/ANSQ Advanced Targeting Forward-Looking Infrared Camera (FLIR) mounted on an F/A-18F fighter jet which came in close proximity to one of these objects. All of this data was thoroughly analyzed in a document entitled *A Forensic Analysis of Navy Carrier Strike Group Eleven's Encounter with an Anomalous Aerial Vehicle*, published in March of 2019 by the Scientific Coalition for UAP Studies, which is available on the internet.

The tic-tac encounter began on November 10, 2004, when one of the ships comprising the CSG 11, the *USS Princeton*, began monitoring strange objects in the vicinity of the strike group. At the time, the CSG 11 was off the coast of California, training for an upcoming deployment to an active theatre of war in the Middle East. The *USS* Princeton was equipped with a SPY-1 radar system which sends out continuous radar pulses at very high rates of speed in a variety of directions, sweeping the sky, generating a vast amount of continuous information about the contents of the aerospace envelope around it. The SPY-1 system contains a large array of transmitting and receiving elements that allow a high degree of focus and direction and is capable of generating a dense and intricate portrait of information about any objects that it may encounter. This system can track upwards of 100 objects at once. This type of dense portrait of the aerospace is a far cry from the "radar blips" that most people associate with conventional radar technologies.

Starting on November 10, 2004, radar operators on the *USS Princeton*, picked up sets of anomalous objects on the SPY-1. These objects reportedly appeared in waves of 8 to 20 at a time and were noticed descending from approximately 80,000 feet to near sea level in less than a second. The *Forensic Analysis* calculated that the forces involved in such a maneuver by a physical object would be astounding: the maximum velocity would be as high as 104,894 mph, and acceleration rates would be as high as 12,250 g-forces. These are obviously impossible rates of speed for any human technology to attain, or any human being to withstand, and incomprehensible "foo" from the standpoint of conventional physics.

The *USS Princeton* observed these anomalous objects for several days. On November 14, 2004, a group of 14 of these objects was sighted by the *USS Princeton* and confirmed by the other ship in the CSG 11, the *USS Nimitz*, as well as another aircraft. At the time, the CSG 11 was preparing to commence an exercise involving at least 30 aircraft. The *USS Princeton* was concerned that the objects could pose a threat to pilot safety. The *USS Princeton* tasked the aircraft from the *USS Nimitz* with engaging the objects to see what they were. These circumstances are strong indicators that the objects were real, regardless of what they really might be.

The first engagement occurred at approximately 2 pm. Once the pilot reached the site where radar indicated the object could be found, he noticed a strange motion in the water which reminded him of the type of disturbance that a sinking ship might create. But no object was directly sighted. After the first set of jets left the area, a second set of jets arrived at the location at 2:10 pm. The lead jet in the second group was piloted by Commander David Fravor. Once they arrived at the location, Fravor and the other pilots saw a disturbance in the water. As these pilots continued to observe that location they witnessed an anomalous object, the giant tic-tac described above. According to the *Forensic Analysis*, Fravor described the tic-tac as follows,

> CDR Fravor described a white "Tic-Tac" shaped object, with perhaps two small appendages hanging below its belly, moving just above the water disturbance. The object had no wings or exhaust and its movement had no observable effect on the calm ocean surface such as that of a rotor wash from a helicopter. CDR Fravor estimated the object to be 50 feet above the water and he described its movement as follows: "It's almost like a ping pong ball. So when it goes right it can stop instantly, and it goes back left, it goes straight forward, it is randomly moving around, very erratic."

Fravor engaged with the object, which he described as follows,

> So we passed through about the twelve o'clock position and we're descending. It [the "Tic-Tac"] kind of recognizes that

we're there and it starts to mirror us. [The same thought went through the wingman pilot's mind who stated, "The UFO turned on them as if it knew or somehow anticipated what they were going to do."] So now, think of it at the six o'clock position, we're at the twelve o'clock position. We're coming down and it starts coming up. So it's going towards nine o'clock and we're going towards three o'clock. And we do this all the way around until I get all the way back towards about the nine o'clock position. So I'm still coming down nice and easy and I'm watching this thing. Because it's just kind of watching us and following. And I'm like, 'That's kind of weird.' So now there's probably about, let me think, 2,500, it's probably about maybe 3,000 feet below us and about a mile across the circle. It's about the size of an F-18. So you know 47 feet long. But it has no wings. I don't see any exhaust plume, you know, like an older airplane would have smoke. There's none of that.

So as I come across, I'm a little above him. He's at the three o'clock position and I go, 'Well, the only way I might get this is to do an aggressive out-of-play maneuver." So I dump the nose and I go from the nine o'clock through the vertical down, to go across to the three o'clock. So he's over here and I go like this [motions cutting across the circle]. So as I get down to about, I'm probably about 60 degrees nose low a little, pulling through the bottom. It starts to accelerate. It has an incredible rate of acceleration. And it takes off and it goes south. And it takes off like nothing I've ever seen. It literally is one minute it's there and the next minute it's like, poof, and it's gone.

After disappearing from view, the tic-tac re-appeared at a specific location known as a "hold point". The hold point is a location in space composed of predetermined latitude, longitude, and altitude points where fighters station themselves to protect an asset and in this case was set in advance of the exercises. These coordinates were only shared between the pilots and those on board the ship with "a need to know." According to officers on the *USS Princeton*, the tic-tac travelled precisely to the hold point and loitered there long enough for its presence to be noticed and discussed among the service members to their extreme shock and surprise. Due to fuel limitations, the jets soon had to return to the *USS Nimitz* to land.

The first two sets of jets engaged the anomaly at close to the same time. Roughly an hour and a half after this engagement, a third set arrived in the vicinity of the previous activity and re-engaged with the tic-tac. It was during this 3rd encounter that footage of the tic-tac was recorded on a FLIR camera mounted on one of the jets. The FLIR camera is designed to operate in the infrared, meaning that the image of the tic-tac on the camera did not come from optical band of the electromagnetic spectrum. The tracking system on the FLIR system is used to lock onto high-speed objects, such as missiles, and is extremely accurate for these tasks. The footage from the FLIR camera is approximately 76 seconds in length and widely available on the internet. Its provenance has been verified by the U.S. military. The footage captures an oblong object moving about a tracking frame. After moving back and forth, the tic-tac darts away from view, eluding the tracking system on the FLIR camera in a way that was incomprehensible to the pilots. The high rate of speed observed in the video footage was consistent with the previous observations of these objects from the *USS Princeton*.

After the engagement, Fravor made copies of the FLIR video and watched it several times. The engagement was a topic of discussion among the members of the CSG 11. Observers on the *USS Princeton* had a bird's eye view of the engagements through the Command Information Center (CIC) on that ship. This CIC system collected data from all of the jets, and also the data from the SPY-1 system. Observers on the *USS Princeton* reported seeing multiple other objects react during the time that Fravor engaged with the tic-tac. Observers on the *USS Princeton* were extremely excited by the interaction. The events witnessed through the *USS Princeton's* monitoring systems were described by one observer as follows,

> At a certain point there ended up being multiple objects that we were tracking. That was towards the end of the encounter and they all generally zoomed around at ridiculous speeds, and angles, and trajectories and then eventually they all bugged out faster than our radars. We were getting what we call 'spot radar sightings' where it would just catch a glimpse of it as it was moving so it was moving faster than our radar could register. And then they were gone.

The engagements generated a tremendous amount of data on the computer recording devices for the ships and jets. Immediately after the event, witnesses said unknown individuals landed on the ships comprising the CSG-11 and confiscated nearly all data related to this encounter. It is presumed that these individuals were from other departments in the U.S. government. In subsequent requests for information regarding this incident, the U.S. government denied that any data was missing or had been taken. Only the tic-tac video and other routine data from the Navy vessels were obtained.

The tic-tac encounter is by far the most detailed and well-documented engagement with this class of anomaly in human history. In addition to the multiple eye-witness reports from well-trained observers, there is additional circumstantial and direct evidence which corroborates many of the specific details of the occurrence. Moreover, the length of the encounter—spanning over several days—is completely unprecedented in the annals of publicly-accessible sightings. Because of this, scientists from the SCU were able to provide reliable calculations concerning the velocity and acceleration of the objects even though the lion's share of the data was taken or destroyed by unknown government officials. A fair amount of direct data was generated during the encounter. One observer on the *USS Princeton* described what further information could have been learned from the missing data as follows, "you could literally plot the entire course of the object, you could extract the densities, the speeds, the way that it moved, the way it displaced the air, its radar cross-section, how much of the radar itself was reflected off its surface. I mean you could pretty much recreate the entire event with the CEC data." The tic-tac encounter demonstrates that it is currently within the technical capabilities of humans to study these types of anomalies in great detail.

2.11.2 *Lesser-known anomalies*

While the tic-tac encounter is the most detailed and well-observed of this class of anomaly available to the public domain, it is by no means the only one. For instance, in January of 2015 a U.S. Navy pilot in an F/A-18F fighter engaged with an object that looked like a gimble off the Atlantic coast and displayed flight characteristics which could not be accounted for. The object was also accompanied by a fleet of these objects, which the pilot reported seeing on his radar as well. A 34-second clip of this video was released with the New York Times article and confirmed as authentic by the U.S military. As with the tic tac, the observed mechanics of this anomaly have been studied and analyzed. No known aerospace technologies are capable of the movements this anomaly displayed.

A lesser-known example is the so-called "rubber duck" anomaly. This anomaly was captured by FLIR technology on November 23, 2019. In this case, the footage was taken on jet being flown by the Arizona National Guard on a monitoring mission of the Arizona/Mexico border. The video contains a 30-minute continuous recording of this anomaly, which appears to look like a rubber duck cruising through the air. The anomaly was analyzed by the SCU in a report entitled *A Forensic Analysis of "Rubber Duck," a Publicly Available Video Showing a UAP Purportedly Filmed by Arizona National Guard on November 23, 2019.* After an exhaustive and detailed analysis of all of the possible causes of the anomaly, the report concluded that "the object cannot be explained as any known natural phenomenon or by any known technology used for aeronautical propulsion."

Encounters such as those involving the tic tac, gimble, and rubber duck, likely represent the tip of the proverbial iceberg when it comes to the types of anomalies that are occurring in earth's aerospace. Most of us, of course, won't see them, just like we do not see the workings of the atom directly, but rather only its effects. In order to gain visibility into this new horizon of knowledge, it is necessary to put the same level of commitment to studying these anomalies as we have invested in studying the atom, subatomic particles, viruses, or any other object whose presentation is concealed from our sight at first.

2.12 *Alien anomalies and scientific understanding*

One of the more interesting anomalies concerning this subject matter is why incidents such as the tic tac encounter, as well as many others, have not ignited a revolutionary inquiry within the field of conventional thought, and excited enough members of the human species to undertake a serious collective effort to understand them. In other words, why have we gotten so worked up about the real possibilities of things like atoms, quarks, viruses and black holes, but not actively-present aliens? While we could delve into this question in great, searching detail in order to unearth exacting conceptual answers, we don't need to do so in order to identify the existence of the obstacles themselves with enough precision to understand what we need to do to move forward.

Regardless of the real reasons for the lack of real interest in this subject, we can see this is a collective inhibition. It appears to be similar to the same type of conundrum that presented in the Era of Copernicus. In that case, even though there were people who knew better and understood that Copernicus was probably right, they still just couldn't go there and admit to themselves and each other that, yes, it is true: the earth probably does actually move in space.

In Ezekiel's time, alien anomalies would be automatically classed as the result of a supernatural cause because there would not be any other paradigms within which the phenomenon could be described or understood. But that is not the case today, as we do have the category of "extraterrestrial" or "intelligent alien" which is a plausible, even though highly exotic, class of ultimate cause. However, conventional science categorizes the causes of these anomalies as "unknown" or "unidentified"—just like people in the Era of Copernicus refused to admit that the earth probably really moved in space. This categorization creates an impression that it is impossible to ever discern the real, ultimate causes for the anomaly; it is an expression of subconscious resistance, not real knowledge.

Every real phenomenon must have a real cause. If the circumstances suggest that the cause of an anomaly is not natural, and not the result of human error, delusion, or secret human technology, then the one remaining cause that has not been ruled out (i.e., an "alien cause") remains plausible from a standpoint of modern thought. Instead of using the identifier "unknown" or "unidentified", anomalies of this nature are more accurately described as "alien" if it appears from the direct and circumstantial evidence that the anomaly likely has a real cause and is not simply the product of human error or delusion, or supernatural forces.[13] In

---

[13] From the standpoint of science, saying something has a "supernatural" cause is logically equivalent to saying it has "no cause" since science does not study supernatural occurrences. However, in a strict etymological sense, we could say that an "alien" cause is "supernatural" because it is not describable within the paradigms of either natural or human orders,

modern English usage the term "alien" derives from the Latin word *alienus*, which can mean unfamiliar, strange, suspicious, or simply "other". On a superficial level, usage of this term means that the anomaly cannot be immediately explained by reference to any known paradigm of science, and that the mind does not naturally see a true cause for it. However, the term "alien anomaly" also implies that an intelligent, extraterrestrial being ***could have been*** the actual cause of the anomaly based on the characteristics it displayed; namely, an anomaly that possesses a technological character, which seems to be the purposeful result of a design based on a highly advanced understanding of physics that is beyond the present understanding of humans.

This is not to say that the suggestion of a superior alien intelligence by use of such a term "proves" anything scientifically; nor is this is to say that there exist paradigms within which this anomaly can be understood. For as we shall see in the course of this work, in cases of major paradigm shifts there is sometimes as substantial "lag time" between the moment when the phenomenon is determined to be "real", and the time it becomes a proper object of scientific thought and examination. *See* ***Section 4.6*** ("scientific conversion and the 'discovery' of oxygen"). Instead, it is simply a direct acknowledgement that of the conventional categories of causes for the anomaly, an intelligent alien presence from another planet does naturally come to mind as an initial impression, and that the other categories or causes are inadequate to account for the phenomenon at the present time. If aliens were not a possible cause of some of these anomalies, and no other conventional cause existed, one would simply say the phenomenon did not occur, was the product of human perceptual error or delusion, or a visit from a supernatural being. But if one cannot say that the phenomenon did not happen, and cannot say with reasonably certainty that it was the product of human error, delusion, or natural forces, then the only remaining cause left for further examination and discussion is "alien" in nature. This is very different from

---

and stands outside of them. But calling them "supernatural anomalies" would not have any real meaning for science; by contrast, calling them alien anomalies would because there is a scientific basis upon which we could conclude that aliens, and alien technologies, are real even though they are not a standard feature of the terrestrial order with which we are acquainted. Also, we may presume that at least some of the alien beings are biological in nature and can therefore be considered in that kind of way.

describing the anomaly as "unknown" or "unidentified" which implies that there is no ultimate cause for the anomaly and that there is no real plan to ever determine its ultimate cause.

2.13 *Extraordinary scientific investigations*

Discarding the conventional use of the terms "unknown" or "unidentified" and replacing their use with multi-purpose identifier "alien", serves an important and necessary purpose in shaping and forming the field of collective thought, and is not simply a semantic endeavor. History teaches that, for certain types of anomalies, a real solution must be assumed as predicate to actually finding one. In some instances, a paradigm that explains a given anomaly cannot come into focus unless the possibility of a rational description of the anomaly is somewhat defined in advance of the discovery. Thus, an *a priori* belief in a particular solution allows the process of defining and articulating a paradigm to begin, and without that belief a future articulation is simply not possible. In order for a study of the anomaly to begin in earnest, it is necessary to get out of the phase of novelty suppression and move into what Kuhn refers to as the phase of "extraordinary investigations".

> Normal science, the activity in which most scientists inevitably spend almost all their time, is predicated on the assumption that the scientific community knows what the world is like. Much of the success of the enterprise derives from the community's willingness to defend that assumption, if necessary at considerable cost. Normal science, for example, often suppresses fundamental novelties because they are necessarily subversive of its basic commitments. Nevertheless, so long as those commitments retain an element of the arbitrary, the very nature of normal research ensures that novelty shall not be suppressed for very long. Sometimes a normal problem, one that ought to be solvable by known rules and procedures, resists the reiterated onslaught of the ablest members of the group within whose competence it falls. On other occasions a piece of equipment designed and constructed for the purpose of normal research fails to perform in the anticipated manner, revealing an anomaly that cannot, despite repeated effort, be aligned with professional expectations. In these and other ways besides, normal science repeatedly goes astray. **And when it does—when, that is, the profession can no longer evade**

> **anomalies that subvert the existing tradition of scientific practice—then begin the extraordinary investigations that lead the profession at last to a new set of commitments, a new basis for the practice of science.** The extraordinary episodes in which that shift of professional commitments occurs are the ones known as . . . scientific revolutions. They are the tradition-shattering complements to the tradition-bound activity of normal science. (emphasis supplied)

Examining this same type of issue in a different context, Bohm refers to it as the "extraphysical principle" in *Causality and Chance in Modern Physics* (1957), as follows:

> [T]he history of scientific research is full of examples in which it was very fruitful indeed to assume that certain objects or elements might be real, long before any procedures were known that would permit them to be observed directly. The atomic theory is just such an example. For the possibility of the actual existence of individual atoms was first postulated in order to explain various large-scale regularities, such as the laws of chemical combination, the gas laws, etc. On the other hand, it was of course possible to treat these large-scale regularities directly in terms of macroscopic concepts alone, without the need for the introduction of atoms. Certain nineteenth-century positivists (notably Mach) therefore insisted on purely philosophical grounds that the concept of atoms was meaningless and even "nonsensical" because it was not then possible to observe them as such. Nevertheless, evidence for the existence of individual atoms was eventually discovered by people who took the atomic hypothesis seriously enough to suppose that individual atoms might exist, even though no one had actually observed them.[14]

The conventional perspective (which we will examine in greater detail below in the analysis of the "SETI" paradigm) is that aliens are not visiting, and have not visited, earth. Conventional thought

---

[14] Another example where science assumed a truth in advance of finding evidence to support it is found in the development of conservation of energy theories in the 19th century. *See Kuhn,* "Energy Conservation", in *Essential Tension: Selected Studies in Scientific Tradition and Change* (1977).

assumes that an extraterrestrial alien would behave like people see in the movies or otherwise according to religiously derived fantasies and would make their presence widely and visibly known if they were here. Conventional thought assumes that aliens would be very excited to be our friends and teach us the secrets of their technologies and save us. Because of this collective childish fantasy, when alien anomalies refuse to perform in alignment with these expectations, conventional thought either suppresses them, disregards them, or otherwise fantasizes about them, wishing they would show themselves in some other way, gazing up at the sky like a subterranean homesick alien wondering aloud, "Are we alone?"

This is the same conventional attitude that would have grudgingly admitted that Copernicus' calculations were right, but refused to admit the facts making them possible were probably true as well. Stated differently, this is an example of a collective neurosis that arises at moments in human history where paradigms shifts of revolutionary magnitude are afoot. Because this doesn't seem to happen very often, we are generally caught off guard and mired in the quicksands of ignorance and pure negation when it arises

2.14 *Lessons from the Copernican Revolution*

Copernicus was forced to assume the earth moved about the sun in order to begin to make his observations. But before he began his work, he had no data to support his assumption. Moreover, his assumption appeared patently absurd, insane, and ridiculous from the standpoint of common experience in the 16th and 17th centuries (which is how many people react to the possibility of aliens visiting earth today). But if Copernicus had simply succumbed to that "insane" assumption from the start, there would be no subsequent observations at all. So, then, what prompted Copernicus to disregard the received experience of his time and press on in spite of all the daunting obstacles that lay before him? Why didn't he just give up and go with the flow of his time and surrender to the limits of conventional thought?

In a prefatory letter that Copernicus sent to Pope Paul III with the publication of *De Revolutionibus,* he explains the reasons that motivated him to take this leap of scientific faith. Copernicus identified the lack of mathematical certainty regarding the orbits of the Sun and Moon under the earth-centered model, and the use of *ad hoc* and inconsistent "principles and hypotheses" to explain the motions of celestial bodies, as rational reasons for moving the analysis forward in the absence of direct proof of his assumptions. Copernicus was also critical of the fact that proponents of the earth-centered model failed to provide a coherent account of the structure of the universe itself. As stated by Copernicus on this last point,

> With them it is as though an artist were to gather the hands, feet, head, and other members for his images from diverse models, each part excellently drawn, but not related to a single body, and since they in no way match each other, the result would be monster rather than man.[15]

In other words, Copernicus sensed that something just wasn't right, something was missing, and that a new direction was needed in order to understand the situation. While at bottom this was nothing more than a hunch, an intuition, an informed guess, it is clear that it nonetheless had very powerful effects and was critical in setting the stage for what would come next.

But although a change in thought is a necessary prelude towards the development of new forms of knowledge, it is not, by itself, sufficient. Copernicus' work alone, for example, did not cause the Copernican Revolution although it did lay the groundwork for it to ripen in the field of thought. According to Kuhn, it was Galileo's telescope that ultimately provided the sufficient cause.

Prior to the arrival of the telescope, the implications of Copernicus' work were largely theoretical and confined to esoteric debates. However, once people could see for themselves the moons of Jupiter and other celestial objects through the telescope, the implications of Copernicus' work became real. Embodied, one

---

[15] Bohm notes that the use of increasingly complex epicycles to make the Aristotelian-Ptolemaic cosmological model function also called into question its underlying premise that there was a fundamental distinction between "earthly" matter and "heavenly" matter, which had dominated in the Aristotelian model and played a significant role in the resistance to Copernicus' ideas. "[W]hen the number of epicycles began to increase greatly, one should have begun to suspect that the distinction between heavenly matter and earthly matter was not a fundamental one. But for various reasons (religious, political, psychological, etc.) this notion was not seriously entertained for a long time."

might even say. Since the ideas of Copernicus were at variance with the facts of ordinary life, and it would be impossible for people to directly observe the earth moving around the sun, a direct experience of seeing the moons of Jupiter seems to have been necessary for people to understand that Copernicus wasn't making this all up. At that point, people could no longer evade the anomaly and were forced to confront it. Once this phase was reached, the suppression was confronted and resolved (subsequent, of course, to Galileo's conviction) thus permitting a new phase of study to unfold.

2.15 *Other forms of scientific revolutions*

But knowledge revolutions happen in different ways, and the Copernican-Galileo example is merely one example among many. For instance, Newton's *Principia Mathematica* is widely regarded as one of the most revolutionary works in the history of thought. Kuhn says the *Principia* marks a decisive turning point in scientific inquiry from which we have never looked back. But Newton's revolution was welcomed, and the anecdote of the falling apple and the formulation of the universal law of gravity is a celebrated example of scientific insight. The reason why this might have been the case was because a vacuum had been created from the time of Copernicus-Galileo which was begging to be filled with a comprehensive theory that would make people feel at home again. Newton's narrative of universal laws perhaps provided a much-needed sense of stability by making the universe somewhat predictable and steady and permitting it to fit together again.

Kuhn notes that paradigm-shattering scientific revolutions usually begin with a period of acute distress where the familiar landmarks and paradigms no longer seem adequate. The landscape begins to look strange and foreign. This distress is occasioned when anomalies are encountered which do not seem to fit into any of the systems that have been developed, disrupting the sense that we understand what is going on in our world. Examples of other types of anomalies include problems faced by chemistry and electricity in the 18th and 19th centuries.

2.15.1 *The electrical revolution*

In certain respects, the problems presented by electrical research are very similar to those presented by alien anomalies. Electricity has always been an observable aerospace phenomenon, but until the 18th century it remained beyond the reach of humans and was historically attributed to the actions of supernatural beings. In an essay entitled, "Mathematical versus Experimental Traditions", Kuhn states the scientific study of electricity "was spawned by efforts to find the relation of the magnet's attraction for iron to that of rubber amber for chaff". When "electricians" as they called themselves, began to seriously study this phenomenon, there was an extended period of time during which there were numerous views about the nature of electricity but no one theory or paradigm that could universally explain the phenomenon. Kuhn says these various schools and theories had a family resemblance to each other, but nothing more than that. For instance, some theorists thought attraction and frictional generation were the most fundamental aspect of the phenomenon and treated repulsion as a secondary effect. Another group thought that attraction and repulsion were equally elementary manifestations. A third group regarded electricity as a "fluid" that ran through conductors, rather than an effluvium that "emanated" from non-conductors. Research into electrical activity in its early stage was more akin to a random fact-gathering activity, until a paradigm came into focus which ended many of the debates among these schools, settled fundamentals, and gave scientists the sense that they were on the right track. As a result of this confidence, scientists were able to undertake "more precise, esoteric and consuming sorts of work," Kuhn tells us. This enabled the theorists to "pursue selected phenomena in far more detail, designing much special equipment for the task and employing it more stubbornly and systematically than electricians had done before. Both fact collection and theory articulation became highly directed activities". However, if the electrical researchers of the 18th century had never posited any causes or characteristics for the observed phenomena and were content to simply say that the causes of electricity were "unknown" or "unidentified" then clearly no forward progress would have ever been made.

2.15.2 *The X-ray*

Another example of a similar type of anomaly is the presentation of the X-ray. According to Kuhn, the discovery of the X-ray was unexpected and came out of nowhere. On November 8, 1895, German scientist Wilhelm Roentgen had been conducting a normal investigation of cathode rays in his lab when he noticed that one of his other pieces of laboratory equipment was glowing while he was examining the rays. After investigating the matter for 7 weeks, Roentgen determined that the equipment was glowing as a result of some form of radiation from the cathode ray tube and that the agent of the strange glow had some similarity to light. When he revealed his findings, Lord Kelvin "first pronounced them an elaborate hoax," Kuhn tells us.

Even though the dramatic appearance of the X-ray in 1895 was bizarre and strange and not predicted by any theories or paradigms, it did not cause any serious scientific upheaval and even Lord Kelvin's esteemed resistance was ultimately surmounted. This was because although the X-ray was not anticipated, it was also not in deep conflict with any existing paradigms. As Kuhn explains,

> To be sure, the paradigms subscribed to by Roentgen and his contemporaries could not have been used to predict X-rays. . . . But neither did those paradigms, at least in any obvious sense, prohibit the existence of X-rays as the phlogiston theory had prohibited Lavoisier's interpretation of Priestly's gas. On the contrary, in 1895 accepted scientific theory and practice admitted a number of forms of radiation—visible, infrared, ultraviolet. Why could not X-rays have been accepted as just one more form of a well-known class of natural phenomena? Why were they not, for example, received in the same way as the discovery of an additional chemical element?

Kuhn suggests that the reason why X-rays were so shocking was due to the fact that they violated "deeply entrenched expectations" concerning "the design and interpretation of established laboratory procedures." Kuhn notes that scientists had been using cathode ray equipment all over Europe in the 1890's, yet none of them had known that these cathode rays were also emitting X-rays. This caused scientists then to suddenly doubt themselves in an uncomfortable way. As Kuhn puts it,

> If Roentgen's apparatus had produced X-rays, then a number of other experimentalists must for some time have been producing those rays without knowing it. Perhaps those rays, which might well have other unacknowledged sources too, were implicated in behavior previously explained without

reference to them. At the very least, several sorts of long familiar apparatus would in the future have to be shielded with lead. Previously completed work on normal projects would now have to be done again because earlier scientists had failed to recognize and control a relevant variable. X-rays, to be sure, opened up a new field and thus added to the potential domain of normal science. But they also, and this is now the more important point, changed fields that had already existed. In the process they denied previously paradigmatic types of instrumentation their right to that title.

2.16 *The framework of scientific revolutions and alien anomalies*

These sparse examples show that scientific anomalies take different forms and arouse conventional resistance in different ways. The woman on the street in 1895 was probably not nearly as disturbed by the discovery of the X-ray as the European scientists were and may have actually found the revelation of the X-ray to be quite stimulating. According to Kuhn's analysis, such a reaction is possible because a person outside of the closed segments of scientific specialists is not personally invested in a particular theory being right or wrong. Moreover, since they themselves are not conducting investigations and placing enormous time and effort into pursuing the correctness of particular theories, they have no personal reasons to be disappointed by the emergence of the X-ray in the way that a person like Lord Kelvin was. Indeed, precisely because Lord Kelvin was an esteemed and accomplished scientist with strong views on how things should be, he would have every right to have a strong reaction to the X-ray. But regular people outside of these closed circles of thought could be more interested or amazed by the discovery instead of being disturbed or troubled.

The same thing cannot be said, however, of the Copernican Revolution, in which scientific advancement disrupted the stability of the ordinary facts of life for everyone, and not just within the closed circles of scientific thinkers. By contrast, the development of a paradigm for the understanding of electricity cannot be said to have caused any real disruption at all, since all the scientists involved in this field of study could see that their paradigms were incomplete until a paradigm rose to the fore which explained the phenomena of electricity in a way that made it seem less anomalous and more normal. Once the power of electricity was harnessed, it became an object of great delight and wonder.

These examples shed important light on the particular reasons why new ways of viewing the world sometimes meet fierce resistance and how that resistance can be overcome. In the case of scientific research, while it is necessary to search for what we might call "objective" truth that is not dependent upon the particular viewpoint of the researcher, that drive itself to find this objective truth is ultimately subjective and personal in nature. Thus, if it is thwarted and the scientist lets his or her emotions get the best of them (as happened to Lord Kelvin) they can be led astray. *See* **Section 5.3.5** ("correcting the barriers to paradigm formation").

By considering these examples, we see that the nature of the resistance to new ideas varies based on the context and circumstances. The X-ray resistance could be quickly assimilated; the Copernican resistance took 80 years and Galileo's imprisonment to overcome. Electricity, by contrast, was approached as if it was a mystery from the start, so there was not a resistance to a pre-established paradigm—just a period of very difficult and vexing study.

It would seem that the main reason alien anomalies are resisted is because they seem so unbelievable, so unreal. How could it possibly be that real aliens are visiting earth and our environment in which we can witness them? Just as people 500 years ago found it unreal and unimaginable that the earth could be moving in space because of the long-term conditioning to which they had been subjected, so too can we today generally not even comprehend how aliens could be real and embodied. The mere thought of this, taken seriously, strikes the conventional mind as an absurd possibility on its face. This strangeness, and reluctance to examine, can be overcome, however, if we learn to regard alien anomalies as "pre-scientific phenomena", a matter which is discussed in **Section 5.3.**

It would be one thing if aliens came down to earth in a grand, public display, making sure that everyone saw them at once and couldn't doubt them. For this, indeed, is how the conventional mind expects a supernatural being to appear, tearing apart the fabric of reality in a divine spectacle. This is also how the conventional mind would expect a real alien to appear as well; after all, if they have the technology and capability to travel across the galaxy, certainly they would understand the importance landing on earth in a spectacular way and enlightening us with their alien knowledge and technologies.

This type of conventional certainty appears structurally similar to the mindset that stymied and delayed the Copernican Revolution for so long, and it should be rejected. Instead of believing in self-created, *a priori* certainties, we should take a much more modest and precise approach. All we really know is that we seeing something there, something going on, but know nothing more about it than that. Some of these anomalies naturally suggest to the mind something technological is going on, as opposed to something natural or animal; and yet, this technological manifestation appears to us something from the annals of science fiction itself. Instead of happening in the confines of a movie theatre, it appears to be happening on the stage of reality itself. Adding to the mystery is the fact that these appearances are elusive and fleeting. The behavior is reminiscent of an exotic, rarely-observed species of animal (i.e., "galactic megafauna"), and not the type of calculating, rational being that one would naturally expect to find behind a higher-order technology.

Conventional understanding cannot make sense of these anomalies and thus prefers to ignore them altogether so it can get back to reading out the world that it knows and that is familiar to it. However, conventional understanding exposes itself to hypocrisy on this point because it also admits that alien beings likely exist within the breadth and depth of cosmological time and can really think of no good reason why aliens shouldn't already be present on our planet. It is this inherent tension between the general tenets of conventional understanding, and the lived conventional experience of the day, that provides the space within which new knowledge may emerge. This situation should be exciting, not depressing.

2.17 *How might the potentials of conventional knowledge change if alien anomalies were admitted into study?*

If conventional resistance was overcome and alien anomalies were admitted into study in a manner reminiscent of the study of the earth's position in space, light, X-rays, electricity, and the like, how might things be different? Take as an initial assumption that after

Project Sign closed, its classification status was downgraded as recommended by the report, and that the U.S. Air Force took a more cooperative and curious interest in learning about these anomalies. We know that the Project Sign approach proposed the use of standardized reporting procedures in order to rule out the possibility of prosaic explanations for the anomalies, such as "guided missile research activity, weather and other atmospheric sounding balloon launchings, commercial and military aircraft flights, flights of migratory birds, and other considerations . . .". We also know that this approach proposes that a "technical analysis [be] made of some of the reports to determine the aerodynamic, propulsion, and control features that would be required for the object to perform as described in the reports". This open approach is described in the Project Sign document as follows,

> The possibility that some of the incidents may represent technical developments far in advance of knowledge available to engineers and scientists of this country has been considered. No facts are available to personnel at this Command that will permit an objective assessment of this possibility. All information so far presented on the possible existence of space ships from another planet or of aircraft propelled by an advance type of atomic power plant have been largely conjecture . . .
>
> Many sightings by qualified and apparently reliable witnesses have been reported. However, each incident has unsatisfactory features, such as shortness of time under observation, distance from observer, vagueness of description or photographs, inconsistencies between individual observers, and lack of descriptive data, that prevents definitive conclusions from being drawn. Explanations of some of the incidents revealed the existence of simply and easily understandable causes, so that there is a possibility that enough incidents can be solved to eliminate or greatly reduce the mystery associated with these occurrences.
>
> Evaluation of reports of unidentified objects is a necessary activity of military intelligence agencies. Such sightings are inevitable, and under wartime conditions rapid and convincing solutions of such occurrences are necessary to maintain morale of military and civilian personnel. In this respect, it is

considered that the establishment of procedures and training of personnel is in itself worth the effort expended on this project.

So how might things in the real world of today be different if the military had been engaging in the Project Sign approach since 1947, instead of the one that it actually did?[16] If the Project Sign approach had been adopted, one can imagine, for instance, that the Marianas film could have been handled in a much different way. One can imagine that when Marianas sent his film to the Air Force, it might have been seriously analyzed and considered. Under this scenario, the Air Force might have returned the film to Marianas in the same condition it was when it received it, and would not have altered or destroyed it. Perhaps when the film was returned to Marianas, the Air Force would have candidly and honestly shared its view of what was observed on the film. Perhaps the Air Force would have used its institutional prestige and resources to solicit the considered views of scientists and other experts. While the Air Force would surely have to keep some of its work product and analysis under classification for national security purposes, having recognized that these alien anomalies pose no direct threat to national security, it would presumably be freer to discuss them with the general public.[17] Maybe the Air Force would have even provided Marianas with a letter of acknowledgment, stating in general terms that it had the film reviewed by experts, concluded that it was genuine, and does not have any explanation for the anomaly that was recorded.

---

[16] In this connection, let us make the following observation. Due purely to historical accident, the U.S. military has been the entity with the best viewpoint to observe these alien anomalies. In a sense, one could say that the U.S. military has played the role of "Galileo's telescope" when it comes the subject matter of alien anomalies. However, unlike Galileo's telescope, which was open for public viewing in the interests of democratizing scientific knowledge, the U.S. military has consistently blocked the public from seeing through the lens based on fear, ignorance, and the irrational superstition that humans are incapable of accepting the fact that they share the universe with intelligent alien life.

[17] Law enforcement agencies sometimes solicit the help of the general public in solving crimes. When they do so, they release some of the details but keep others confidential for purposes of their ongoing investigation.

Under this imagined scenario, the Marianas film could have had a far different impact on the knowledge potentials going forward. Armed with an acknowledgement from the U.S. Air Force that the film was authentic, and that the anomaly could not be explained, Marianas would then be free to exercise his rights under the First Amendment to show the film to others, and perhaps himself solicit the viewpoints of other scientists on the matter. Maybe the scientific community would take a more receptive attitude to a film whose authenticity was verified, and whose contents were not understood. Perhaps the general public would be less inclined to ridicule Marianas for showing the film, and more inclined to listen to him. One could only imagine how different things might be if millions of Americans were drawn to the same stunned silence upon viewing the film that the members of the Great Falls Lion's Club were in 1951.

This thought experiment shows us how the conceptual landscape might change when novelty suppression is removed. Had this approach been taken in 1950, an anomaly such as the tic tac could have been treated far differently in 2004. By that point in time, an anomaly such as the tic tac could be viewed with great interest, instead of derision or suppression. Perhaps when the CSG-11 first encountered the loitering tic tacs there would even been a scientific task force, well-equipped with a high range of technologies, experts, and specialists, ready to engage the anomaly on a scientific basis instead of a military one. One can imagine how an attitude of unhurried curiosity and interest in the tic tacs could have transformed the engagement into something far deeper than it actually was. For example, if a "scientific version" of the *USS Princeton* was leading the engagement, the initial observation of the tic tacs descending from 80,000 feet to sea level would have been more patiently and thoroughly observed for *at least several days*. The data and information that humanity could have gained from such an encounter would be hard to fathom.

It is also possible that this gathering of tic tacs could have been approached in a more cautious and methodological way, instead of as a threat which needed to be dispelled by the Navy pilots in order to continue their preparations for war. In any event, since the scientific version of the *USS Princeton* would not be in a hurry and would have had different objectives in mind, it could handle the engagement in a completely different manner, in a way that was designed to maximize the knowledge potential of the situation at hand.[18]

The suggestion in this thought experiment can be restated as follows: there is a real relationship between the knowledge gained about these anomalies and the attitude and preparedness brought to bear on their study. If the attitude and approach changes, so do the possibilities of actual knowledge; by contrast, if these anomalies are suppressed, the knowledge potentials remain near zero. This is the actual case in which we find ourselves, where our knowledge of these anomalies in 2025, differs little from the knowledge that Jung described in 1958. Although one cannot say exactly "what" or "how" the state of knowledge will change once the suppression is lifted, the examples of electricity, X-rays, general astronomy, and other similar objects, show that new forms of knowledge are possible if anomalies are admitted into study. This, of course, takes a new kind of effort. The energy to generate the level of commitment necessary will never arise so long as everyone is unsure if there is anything to find out or learn.[19]

---

[18] In particular, an ability to make more precise observations, with equipment which was prepared in advance and designed for the detection of alien anomalies, would lead to the possibility of generating more accurate and reliable measurements for the speeds, velocities, and flight characteristics of alien anomalies. As Kuhn has argued, quantitative anomalies, in the form of measurements which cannot be reasonably disputed, are critical in overturning outdated paradigms and resolving scientific crises. "No crisis is . . . so hard to suppress as one that derives from a quantitative anomaly that has resisted all the usual efforts at reconciliation. Once the relevant measurements have been stabilized and the theoretical approximations fully investigated, a quantitative discrepancy proves persistently obtrusive to a degree that few qualitative anomalies can match. By their very nature, qualitative anomalies usually suggest *ad hoc* modifications of theory that will disguise them, and once those modifications have been suggested there is little way of telling whether they are 'good enough'. An established quantitative anomaly, in contrast, usually suggests nothing except trouble, but at its best it provides a razor-sharp instrument for judging the adequacy of proposed solutions."

[19] It is true that there a number of people who are "believers" in the phenomena of alien anomalies, but that is not what we are discussing here. The intensity of personal belief is one thing, and indeed an important thing, when it comes to the motivation necessary to make individual discoveries. But on a collective level, there is a different type of normal operation that must occur through a process of scientific conviction, or scientific faith, that the search is worth pursuing even though one is unsure

For these reasons, it is necessary to abandon the current convention of using the terms "unknown" or "unidentified" to describe these anomalies, as it inhibits the development of a positive discourse of knowledge and the formation of paradigms of active meaning. Instead, it is necessary to use a productive term such as "alien" to describe these anomalies, thereby situating them within a positive network of meaning in which their highly exotic and extraordinary characteristics can be thoughtfully considered and actually explored.

2.18 *Conclusion*

Human beings in the 21st century truly have an unprecedented chance to understand alien anomalies in a clear and precise way; however, we have not yet taken advantage of this opportunity. The lack of initiative stems from the confusion that naturally arises when new orders of thought clash and conflict. This is the lesson we draw from a consideration of the obstacles that were overcome in the Copernican Revision.

While the technological capability is present, collective will and understanding is lacking; thus, knowledge in this field has been stagnant for over 60 years. There continues to exist a fierce and persistent conventional resistance to admitting that aliens could actually be present around earth today; this collective resistance inhibits the development of genuine knowledge. In order to change this state of affairs, we must change the ways that we think and speak about these things. In this specific case, this requires us to adopt the term "alien" to describe these anomalies in an appropriate circumstance. While it may sound silly or trivial to some people, such a simple shift in language and thought can have enormous repercussions for this field.

---

of success. *See* **Section 7.4** ("Cocconi and Morrison's revolutionary vision").

3.0 SAFE DISTANCE PARADIGM

3.1 *Conventional knowledge and the suppression of anomaly*

Since aliens have never been a part of the common story of planet earth, the possibility of actual human to alien direct contact is generally viewed as "extraordinary" and fictional in nature. And because there has never been a serious scientific investigation into the possibilities of alien contact here on earth, the potential of it occurring has been confined to the realms of fantasy, metaphysical or spiritual speculation, and idiopathic experience.

One can reasonably state that since the 1950's and 1960's, aliens have only been real for the overwhelming majority of people within the confines of science fiction fantasy, likely resulting in a large-scale subconsciously conditioned belief that an intelligent alien and a human could only have direct, real-time contact as a *fictional possibility* instead of a real one. In fact, even when conventional scientists do attempt to think about alien possibilities, their thoughts are soaked in untestable, science fiction fantasy notions such as "Dyson spheres" or deep space "alien megastructures". Adding to these challenges, the idea of an alien visiting planet earth clashes with preconceived ideas about the possibilities of space travel, given the impossibly large distances relative to the present capabilities of human beings. So, for instance, even if faster-than-light travel is possible for alien beings, or even if it is possible to travel great distances over time within the framework established by relativity theory, these distances are still outside of the common experience of human beings; again, faster-than-light and "hyperspace" travel are things that we only see in the movies, not in real life.

Accordingly, the possibility of real alien presence on earth is naturally subject to various forms of suppression and distortion within the human mind. Suppression arises when new observations, hypotheses, or anomalies, conflict with the predictions and express or implied limitations of a dominant paradigm. According to Kuhn, there is a process of discrimination through which we determine when to suppress novelties or anomalies, and when to admit them into study, which occurs *prior* the moment of presentation. The determination of admissibility is made on a case-by-case basis, although the process does appear to follow a generalized structure. Sometimes anomalies are completely rejected and labeled "unknown" or "unidentified", thus ending any form of inquiry. This is what happened to the Copernican inquiry for quite some time and also afflicts the inquiry into alien anomalies in 2025. Perhaps one could say that this sort of thing happens when the consequences of accepting the reality of the anomaly is simply too much for the conventional human mind to bear.

On some occasions, the anomaly is admitted into study and is ultimately explained by a paradigm. X-rays are a good example of paradigm "absorption" because this anomaly was never truly rejected and was ultimately explained by the development or extension of existing paradigms. X-rays were anomalous until they were understood, and an understanding of this new phenomenon constituted an unforeseen extension of a paradigm regarding electromagnetic phenomena which was already in existence. However, there was no irreconcilable conflict between the paradigm of electro-magnetic radiation and the x-ray; while the paradigm did not *predict* the x-ray explicitly, nor did it forbid it. Consequently, it was possible for the anomaly to be quickly absorbed by Roentgen after he locked himself in the lab for seven weeks and shut out the rest of the world. Obviously, if Roentgen had ignored the anomaly, or just classified it as "unidentified", nothing new would have been learned.

Although novelty suppression can inhibit the development of knowledge, it also serves an important function. Suppression results in a restricted scientific vision which allows a narrow focus and increase in commitment to a particular path of study which is not possible if all facts are viewed as equal. "By focusing attention upon a small range of relatively esoteric problems," Kuhn states, "the paradigm forces scientists to investigate some part of nature in a detail and depth that would otherwise be unimaginable . . . during the period when the paradigm is successful, the profession will have solved problems that its members could scarcely have imagined and would never have undertaken without commitment to the paradigm." If suppression of an anomaly is occurring on non-scientific grounds (as was the case with the Copernican resistance), the potential for real knowledge of that anomaly will remain at zero since the suppression stems from arbitrary, non-scientific factors that are merely the expressions of personal inhibitions. The exploration of these inhibitions might teach us something about the characteristics of the observer, but won't shed any real light on the phenomenon itself. By contrast, if the suppression or "restriction of scientific vision" is done in the name of a paradigm that is being articulated or explored, the suppression serves a useful purpose by narrowing the range of attention and focusing scientific efforts.

Cases where the suppression is based on arbitrary, subconscious grounds that are not themselves the product of reasoned deliberation, are difficult to resolve because the suppression is not the result of rational thought or scientific insight, but rather personal biases and prejudices, and thus may be immune to deliberation and reason. In these cases, a dominant paradigm will actively suppress the novelty until the tension becomes too great, and the suppressing paradigm is overruled, obliterated, or modified, leading scientists to convert to the newly emerging paradigm. The Copernican example stands out as an extreme and notable example of this kind of dynamic process. The anomaly which was created by the work of Copernicus effected human society on a number of different levels that transcended the niches of specialized scientific communities, whereas in the case of other scientific revolutions, such as relativity, it was only specialized segments of the scientific community, and not society at large, that felt the immediate impact.[20] Indeed, precisely because the position

---

[20] When Einstein introduced the principle of relativity, the resistance among the scientific community was so strong that there was a book written called "One Hundred Authors Against Einstein". This resistance against Einstein manifested in other ways as well. According to Einstein's chief biographer, Walter Isaacson, after Einstein published his paradigm shattering paper, *On the Electrodynamics of Moving Bodies* (1905), he had great trouble getting a job teaching physics, even though some of the leading lights of the physics community vouched for his work. Indeed, Einstein was forced to continue employment as a patent clerk and pursue other lines of work (such as an inventor and salesman) for several years thereafter. Isaacson relates an amusing tale where in 1908, Einstein applied for a job teaching mathematics at a high school in Zurich, and among 21 applicants, did not even make the list of 3 finalists. It wasn't until February of 1908 that Einstein was finally able to get what amounted to an adjunct professorship, and it was not until March of 1909, a full four years after he revolutionized physics, that he was offered his first full-time job as a physics professor at the University of Zurich (some contend that antisemitism may have also played a role here; if true, it is yet another instance of how arbitrary factors can influence the resistance to, or acceptance of, scientific paradigms).

of the earth in space had become a fact of daily and religious life, arbitrary factors from those domains added special weight to earth's position in space in the minds of humans which was not easily overcome. Conversely, if the fact of the position of the earth's location in space was not important to systems of theology or common understanding, a revision of this fact would be unlikely to generate such resistance. In the case of X-rays, for instance, Lord Kelvin was apparently disturbed by the findings[21], but as far as we know there was no popular resistance to X-rays in common society or among any religious institutions. But if systems of thought or theory were in place which prohibited the extension of the 19th electromagnetic paradigm to new objects such as the x-ray, one could imagine a much greater form of popular or widespread resistance.

Although it might sound absurd to us, imagine a situation where the dominant religious, governmental, or academic institutions had placed their credibility on the completeness of the 19th century electromagnetic paradigm. Imagine that the completeness of the electromagnetic paradigm had become an article of faith in the world at large, instead of simply a grouping of facts studied by a small segment of the social order. In that case, there would be a fierce resistance to the X-ray, instead of simply some offhand comment by Kelvin that was easily overcome. But because this paradigm was mainly operating within closed segments of scientific specialists and was not subject to external social forces, nimble and inquisitive scientists such as Roentgen could easily question its completeness and open the paradigm to new phenomena instead of being forced to destroy it.[22]

---

[21] Kelvin seemed to be the kind of scientist that had very strong ideas about the ultimate nature of things. "During the late nineteenth century, Lord Kelvin, one of the leading theoretical physicists, said that it was no use for young people to go into theoretical physics. He said that the major discoveries of physics were finished, and that what was left was only a matter of refinements and the next decimal points. However, the thing didn't work out that way." Bohm, *Thought as a System* (1994), pg. 101.

[22] This thought experiment allows one to see just how "arbitrary" these articles of faith can be. After all, why was it necessary for the Church to attach such great significance to the perceived centrality of the earth in the universe? Although this was mentioned in the Bible as an important fact, there are many different facts mentioned in that work, and the decision to attach particular emphasis on any one is clearly a discretionary one.

3.2 *Introduction to the "SETI" Paradigm*

In Copernicus' time, the dominant astronomical paradigm was founded on the work of Aristotle who declared that the earth was at the center of the universe, and that all things moved about it. Today, the dominant paradigm with respect to alien anomalies is known as the Search for Extraterrestrial Intelligence, or "SETI". SETI does not investigate any alien anomalies occurring in the aerospace, such as the tic tac encounter described above; nor, for that matter, will it even speak of such anomalies in a serious way. The only types of alien anomalies SETI seeks to investigate are those hypothetically occurring far, far away whose presence may be revealed by deep space transmission of radio signals from alien civilizations. SETI also searches for what it terms "alien megastructures," which are described on page 14 of the publicly available SETI e-book entitled, *10 Places to Search for Life Beyond* (2024), under the chapter entitled "Boyajian's Star," as follows,

> Slightly hotter than our Sun, the star studied by astronomer Tabetha Boyajian had wild and unpredictable variations in brightness. These variations triggered speculations of an extraordinary cause: alien megastructures bigger than planets and constructed to harness the energy output of the star were intermittently blocking the light from reaching our telescopes. Such structures, postulated by physicist Freeman Dyson, fell from favor as an explanation for the star's brightness variations when dust rings fitted the observations more accurately. Still, the idea of observing megastructures is now a mainstream way of searching for intelligent alien life, and Boyajian's Star paved the way.

3.3. *Theoretical foundations of the SETI paradigm considered*

SETI represents the conventional scientific approach to alien anomalies. While the SETI paradigm suppresses any investigation for alien anomalies in the aerospace, it asserts that a "mainstream way of searching for intelligent alien life" is to probe for the existence of "alien megastructures" in deep space. Thus, the SETI paradigm excludes from consideration on *a priori* grounds an entire class of anomalous phenomena before any scientific examination has occurred. An understanding of the basis for the resistance and suppression embodied in the SETI paradigm reveals itself through an examination of the three key theoretical sources that frame and shape the content of its assumptions: (1) Guiseppe Cocconi and Phillip Morrison's paper entitled *Searching for Interstellar Communications* (1959); (2) Fermi's paradox; and (3) Sagan's dictum ("extraordinary claims require extraordinary proof.").

3.3.1 *Aliens as "inferred" objects*

Searching for Interstellar Communications represents the first instance in which modern scientific thought takes the possibility of intelligent alien life as a real one. While there are key aspects of this work that still hold valid today and remain revolutionary in their nature, there are also assumptions guiding this work that appear somewhat arbitrary and are the likely sources of continuing irrational suppression. Cocconi and Morrison begin their work by justifying the search for intelligent alien life in the following terms,

> No theories yet exist which enable a reliable estimate of the probabilities of (1) planet formation; (2) origin of life; (3) evolution of societies possessing advanced scientific capabilities. In the absence of such theories, our environment suggests that stars of the main sequence with a lifetime of many billions of years can possess planets, that of a small set of such planets two (Earth and very probably Mars) support life, that life on one such planet includes a society recently capable of considerable scientific investigation. The lifetime of such societies is not known; but it seems unwarranted to deny that among such societies some might maintain themselves for very long compared to the time of human history, perhaps for times comparable with geological time. It follows then, that near some star rather like the Sun there are civilizations with scientific interests and with technical possibilities much greater than those now available to us.

This is an example of the time-tested practice of positing an *inferred object* whose reality has not yet been directly confirmed by experience itself, but which reason and the factual circumstances suggest is plausibly real. As discussed previously, an ancient example of this kind of thing is the atom, whose existence was hypothesized nearly 2,000 years before its actual physical presence could be verified or even possibly measured. Those ancient scientists who inferred its existence did so solely on the basis of observable facts and the application of reason and logic.

A more recent example of an inferred object from the realm of 19th century science is the "ether". The ether was never directly detected, and now is not considered to be a real object. But in the 19th century, the wavelike property of light logically suggested the existence of a medium of propagation based on common experience. The circumstances known to the scientists at the time provided the impetus to search for this medium. While they did not detect it, they learned valuable information along the way, ultimately setting the stage for Einstein's development of the principles of relativity.

A modern-day version of the search for the ether is the search for "dark matter". Based on cosmological observations, many astronomers believe that there must be some other form of matter which does not directly interact with light that is shaping and forming the observed motions of celestial bodies in the universe (i.e., "dark matter"). They base this belief on reasonable observations of the facts and circumstances which are known to them. While scientists have not yet directly detected this dark matter, they are searching for it with extremely detailed methods of observation in a manner analogous to the scientists of the 19th century who were searching for the ether.

Cocconi and Morrison's deduction that intelligent aliens likely exist elsewhere in the universe based on what we know about our own scientific history and circumstances remains a rational one to this day. While it could ultimately turn out to be the case that human beings are the only scientifically minded beings in the entire observable universe, many consider that possibility unlikely. Since 1959, exoplanetary science, evolutionary biology, and our own technological advancement, make it appear at least plausible that there exist scientifically minded beings elsewhere in the universe. Moreover, our knowledge of evolution informs us to a degree of reasonable certainty that non-scientific beings can evolve into scientific beings through the evolutionary process.

For example, conventional science informs us that humans can trace their evolutionary lineage back to a theoretical being known as "Concestor 13," a nocturnal, non-scientific, shrew-like creature that existed somewhere between 90 and 100 million years ago. If conventional thinking says that humans can evolve from a being like Concestor 13 through the evolutionary process on earth over a period of 100 million years, then it becomes difficult to completely rule out the possibility of a same or similar process happening on another planet within a similar timeframe. Today it is accepted as a given that the universe has been around long enough for evolutionary processes to have likely occurred on countless occasions in places that no human would ever have access to, no matter how long humanity survives. In any event, we have no *a priori* grounds upon which we can conclude that the evolutionary process that we are familiar with is unique to earth and couldn't happen in a similar way somewhere else.

If we grant as at least plausible that scientifically intelligent aliens could have developed on other planets, it also seems logical to infer that if any of these "alien civilizations" actually exist, they could have reached a level of basic scientific maturity far in the past within the vast horizons of cosmological time. When we look at our own scientific trajectory and development, we see how quickly a species acquires a technological ability to transform its relationship to time, space, and matter in rather dramatic ways. Our best scientific understanding of our own history suggests that the use of the wheel, lever, and wedge, matured no more than 15,000 years ago. The earliest known examples of ferrous metallurgy date to 4,000 B.C. in ancient Egypt. The earliest example of written symbolic communication date back to the Cuneiform tablets in Sumer circa 3300 B.C. Accordingly, we may infer that all of these forms of basic technological innovation were at their most nascent stages of development approximately five to six thousand years ago. From the temporal frame of a basic human being, this span of time is epic; from the perspective of cosmological time, it is virtually nothing. If humans can develop our present technology from these "crude" forms to a level of our current sophistication in a matter of 6,000 years, one can imagine how far human technology could advance in another 10,000, 100,000, or 1,000,000 years—a miniscule quantum of time on the cosmological scale. From this, we see how easily the universe can provide sufficient time for an advanced alien civilization to develop that would seem as godlike and unreal to us as we would surely seem to the ancient Egyptians or Sumerians if they could see us today.

3.3.2 *Assumption that aliens are probably looking for us*

Having inferred the existence of alien civilizations in this manner, Morrison and Cocconi then consider the question of how humans might make contact with them and, in doing so, try to "imagine" what these intelligent aliens must be like. Underlying this analysis is the assumption that, if aliens are really "out there", they would be looking for us. "We shall assume," they write, "that long ago they established a channel of communication that would one day become known to us, and that they look forward patiently to the answering signals from the Sun which would make known to them that a new society has entered the community of intelligence. What sort of channel would it be?" Cocconi and Morrison then posit concrete features of this communication channel. They first consider the characteristics of the "optimum channel" which these beings would use, arguing that "[i]nterstellar communication across the galactic plasma without dispersion in direction and flight-time is practical, so far as we know, only with electromagnetic waves. Since the object of those who operate the source is to find a newly evolved society," they write, "we may presume that the channel used will be one that places a minimum burden of frequency and angular discrimination on the detector. Moreover, the channel must not be highly attenuated in space or in the Earth's atmosphere." Relying on these assumptions, Cocconi and Morrison conclude that "the wide radio-band from, say 1 Mc to 10^4 Mc./s, remains as the rational choice" for such beings to use.

Cocconi and Morrison then further assume that the extraterrestrials, in consciously designing their communication, would be mindful of the physical characteristics of the two "backgrounds" against which this transmission would be detected by beings like us: the emission of their own "local star" and the "galactic emission along the line of sight." After examining the characteristics of each, Cocconi and Morrison conclude that 21 cm, the "objective standard of frequency, which must be known to every observer in the universe," would be the ideal frequency at which we should search. "It is reasonable to expect," Cocconi and Morrison tell us, "that sensitive receivers for this frequency will be made at an early stage of the development of radio-astronomy."

Furthermore, Cocconi and Morrison assume that the civilization beaming out the signal would send it "towards all the sun-like stars in the galactic neighbourhood" and that "[t]he support of, say, 100 different beams of the kind we have described does not seem an impossible burden on a society more advanced than our own. Upon detecting one signal," they conclude, "even we would quickly establish mean search beams." After giving due consideration to the "signal location and band-width," Cocconi and Morrison conclude their paper by considering the "nature of the signal and possible sources." They write,

> We expect that the signal will be pulse-modulated with a speed not very fast or very slow compared to a second, on grounds of band-width and rotations. A message is likely to continue for a time measured in years, since no answer can return in any event for some ten years. It will then repeat, from the beginning. Possibly it will contain different types of signals altering throughout the years. For indisputable identification as an artificial signal, one signal might contain, for example, a sequence of small prime numbers of pulses or simple arithmetical sums.

Lastly, they identify the particular stars that we should focus our initial efforts on, based on their distances, direction from our location, and degree to which they would stand out against the galactic background. These include, tau Ceti, O_2 Eridani, ε Eridani, ε Indi, α Centauri, 70 Ophiucus, and 61 Cygni, as well as "all main-sequence dwarfs between perhaps *G*0 and *K*2 with visual magnitudes less than about +6 . . .".

3.3.3. Arbitrary aspects of the SETI paradigm considered

Although the work of Cocconi and Morrison is revolutionary and should be greatly esteemed, time has shown that it suffers from a number of serious defects in its analysis which should incline the skeptical observer to call certain aspects of this paradigm into question. On the positive, or revolutionary, side, we must give Cocconi and Morrison great credit for the bold and imaginative leap of taking the idea of human to alien contact seriously and treating it as a real possibility for our species; for without this leap of of scientific faith, this possibility wouldn't be receiving any serious consideration today. But that doesn't mean their approach is therefore completely free of error.

For example, one strange assumption we see in Cocconi and Morrison's work is the notion that advanced alien civilizations would use the same type of radio technologies as a 1950's scientist. This approach appears on its face to be somewhat contradictory: if these alien civilizations are far in advance of us, then why would we expect them to rely in 1950's human technology, instead of something so exotic as to be virtually unimaginable? Obviously, no rational observer in the early 21st century would still find this assumption convincing.

Another assumption in this work that also appears untenable is the notion that present day alien visitation on planet earth is conclusively ruled out as a real scientific possibility. It is puzzling as to why these scientists so readily suggest a deep space search of aliens, while not also considering that perhaps some of the disturbances in the aerospace were also worth examining as well. While Cocconi and Morrison do not explain why they take this position, obvious reasons come to mind. Scientists like Cocconi and Morrison mainly inhabit laboratory environments and the halls of academia. While they may spend some time gazing through telescopes or looking up at the sky, much of their time is likely spent researching or talking to other scientists in buildings. Individuals such as Cocconi and Morrison probably spent almost no time examining the aerospace for signs of alien anomalies, and their work would never be interrupted by such anomalies. Moreover, they would likely be subject to ridicule if they did seriously search for such things. This is in contrast to members of the Air Force who work in the skies and therefore cannot simply deny the existence of anomalies which interfere with their work. In other words, the actual circumstances in which scientists such as Cocconi and Morrison operate grant them a discretion to evade anomalies which members of the U.S. Air Force cannot; this explains why Project Sign took these anomalies seriously in 1947, while Cocconi and Morrison failed to even note them in 1959.

Additionally, it is worth noting that if aliens are behind any of these alien anomalies, they are certainly not going out of their way to make their presence known in ways that a scientist conditioned by science fiction fantasy would expect. In science fiction movies, aliens make their presence obvious and irrevocable. After all, when people go to watch these movies, they want to see something interesting and exciting, so bringing an alien presence to the forefront is naturally good business practice. So this is perhaps another reason why Cocconi and Morrison do not mention an aerospace survey: since aliens aren't appearing like we see in science fiction books or movies, they most certainly must not be here. But are these science fiction assumptions based on real science?

The answer to this question must be "no". Scientifically speaking, there is no real reason to assume that an alien mind which we know absolutely nothing about would necessarily behave in this way. While it could be one plausible theory of alien behavior, it would be unjustified for us to assume that it constitutes a universal rule of alien behavior. But because scientists have been generally conditioned to believe that all real aliens would behave like the aliens they grew up seeing in movies or reading in science fiction books, when the alien anomalies fail to behave in these pre-conditioned ways, they are likely to just conclude there is nothing there to see or examine. Unless a scientist who is afflicted with this subconscious bias is willing to examine it and overcome it, it will direct his attention away from any anomalies that do not behave in a manner consistent with his science fiction conditioning and he will not be interested in studying them any further, becoming apathetic to their presentation. The difference between scientists and military members comes into play here again. Even if military pilots are also conditioned to believe that aliens would behave in line with science fiction fantasy conditioning, when they see these things in their operating environment on a repeated basis, and the anomalies interfere with their work, they simply cannot be denied. Scientists, however, can chose to ignore these anomalies and continue their work in other ways.

Although Cocconi and Morrison themselves do not make their views on this point explicit, we do see this rationale explained by Dr. Carl Sagan, one of the SETI paradigm's most zealous and virulent proponents, in a C-Span interview, January 8, 1991, as follows,

> On UFOs, I would just say that the evidence . . . No one would be happier than me if we're being visited. It would save me so much trouble, um, but extraordinary claims require extraordinary evidence and I would say that the evidence that we are now or ever have been visited despite the claims of being abducted and, God knows what happens on the spaceship, those claims are crummy the evidence just isn't good. And uh, my mind is open, I'm willing to, to look at new cases but there has been no compelling cases, physical evidence, photographs, a stolen log book of the alien captain, nothing that would convince a skeptic.

This statement is an example of subconscious science fiction fantasy conditioning in operation. In this statement, Sagan's only demand for proof is the type of proof that one would expect to see in a science fiction fantasy movie. When Sagan says he would like to see a "stolen logbook of the alien captain", he is not just being flippant and arrogant but is also expressing the view that only a form of proof which seems like the stuff of science fiction would be sufficient. Sagan does not, for instance, talk about a real way that science might go about investigating whether any of these observed alien anomalies are actually caused by aliens, such as measuring in detail their flight characteristics and attributes. Clearly, Sagan has dismissed the need for any such detailed examination before the examination has even started. He does not suggest that the protocols suggested by Project Sign should be perhaps followed. Instead, Sagan rushes to what he sees are the most outlandish claims (that of alien abductions) and draws an equivalency between the likelihood of *all* those claims being true and the likelihood of *any* aliens visiting earth at any time.

This tendency to be dismissive of the need for a near-earth study of alien anomalies is a prime feature of the SETI paradigm. It is not, however, a product of careful and reasoned deliberation. Accordingly, those who are locked into this mode of thought occupy the same position in relation to alien anomalies as those who resisted the ideas of Copernicus stood in relation to his work in that time.

3.3.3.1 *Fermi's "neurotic" paradox*

At bottom, SETI's exclusive focus on the deep space search is a consequence of its belief that if aliens were here on planet earth, they would reveal themselves to us in ways that have been predicted by science fiction fantasy, making their presence global, irrevocable, and unmistakable, in a revelatory manner. Another expression of this arbitrary position is found in "Fermi's paradox." Although this expression masquerades as "scientific truth", it is really nothing other than an expression of an unproductive neurosis, a secret longing to communicate with aliens that seems impossible to satisfy. It is irrational, arbitrary, and wholly unscientific; those who spend their time contemplating this so-called "paradox" are doing nothing more than giving life to a collective neurosis which will only lead the thinker into an unproductive, dead end. *See* **Section 5.3.1** ("identifying the neurotic side effects of alien anomalies").

In an article entitled *If There Are Aliens Out There, Where Are They?* (Scientific American, January 6, 2016), the authors, Mario Livio and Joseph Silk, state Fermi's paradox as follows,

> Physicist Enrico Fermi famously asked the question 'Where are they?' to express his surprise over the absence of any signs for the existence of other intelligent civilizations in the Milky Way Galaxy. Although many potential resolutions to this so-called 'Fermi paradox' have been suggested over the years, there is still no consensus on which one, if any, is correct. The question of whether we are alone in the Milky Way (or in the universe at large) remains, however, one of the most intriguing questions in science in general, and in astronomy in particular.

Fermi's paradox states that if intelligent aliens do exist, evidence of their presence should be available to the unaided perceptions of humans, requiring no effort at a specialized scientific study. Put differently, Fermi's paradox embraces the assumption that if aliens in deep space would go to inordinate lengths to communicate with us, they would certainly not conceal their presence from us if they made it here but would be motivated to disclose their presence and engage in direct communication and information exchange with us as soon as possible. This is restatement of the view expressed by Cocconi and Morrison that intelligent aliens on other planets would be waiting and watching "patiently" for us to find the channel of communication that they had established far in the past. The difference between Fermi's paradox and Cocconi and Morrison's assumption is that Fermi's paradox directly considers the absence of obvious signs of intelligent alien life in the context of near-earth environment as conclusive proof of the total absence of any sort of alien anomalies; whereas Cocconi and Morrison simply imply that this is the case by their assumption that alien civilizations are looking to contact us, Fermi's paradox makes this point explicit and then becomes a neurotic fixation over the fact that aliens don't seem to be present in the way that the person has imagined. But these viewpoints are simply two sides of the same conceptual coin.

Both views are grounded in the assumed characteristics of alien behavior based solely on science fiction fantasy conditioning.[23]

Based on the assumed characteristics of aliens from science fiction fantasy sources, the SETI scientist has made *a priori* conclusions that constitute the default perceptions of his basic experience and render inadmissible a whole class of evidence (i.e., evidence of alien anomalies in the aerospace environment) before any inquiry has been initiated. Although default perceptions can certainly provide a valid starting point for scientific inquiry, and science fiction itself can be an immensely useful font for such ideas, these perceptions naturally contain a number of untested assumptions that artificially limit the range of understanding.[24] As Bohm explains in *Special Relativity*,

> [W]hile our perceptions do have a subjective side, dependent on the particular background and conditioning of each person, as well as on the general background and conditioning of the whole of humanity, they also have a kind of objective content, which can go beyond this particular and limited background.

---

[23] All of the alien behavior that one sees in science fiction fantasy is based on extrapolation from human behavior; indeed, if it wasn't, no one would relate to it or pay money to see it. Thus, regardless of how many times scientists have been conditioned to believe that aliens want to meet us and would step out of their spaceships and demand to be taken to the White House right away once they landed on earth, if the behavior of alien anomalies is not actually conforming with these expectations, then these *a priori* expectations must be disregarded and, at a minimum, should not serve as an impediment to genuine investigation. While there could be many other reasons why we might choose not to investigate alien anomalies, the fact that their presentation does not conform with science fiction fantasy paradigms should not be one of those reasons.

[24] A paradigmatic example of this kind of thing is the default experience of "simultaneity". In the world of default perceptions, simultaneity seems like an objectively real experience that would be valid in all domains. However, studies in relativity have shown that the experience of simultaneity is a natural illusion created when the speed of light seems instantaneous; when the actual speed of light is taken into account as an integral part of the observational experience and has a real and precise meaning in relation to precise forms of measurement (instead of an imaginary or fantasy meaning that does not translate into real numbers), there is no such thing as a simultaneous occurrence among different observers. Simultaneity is "real" on one level (i.e., the level of natural, or default observation), and "not real" on another level (i.e., the scientific experience) at the same time.

> For the *general* structure of our perceptions (resulting from this background) can be regarded as a kind of *hypothesis*, with the aid of which we approach subsequent experiences in which things have changed not only of their own accord, but also because of our own movements, actions, and probings, which alter our own relationships to our environments. To the extent that the new experiences fit into the continuation of the old structure without contradictions, these hypotheses are effectively confirmed. But if we are alert, we will sense contradictions when they arise . . . When this happens, the brain is sensitive to the discovery of new relationships, leading spontaneously to further hypotheses, which are embodied in the appearance of new structures in the "inner show". Anyone can see this happening as he approaches a distant object that is unknown to him, or as he approaches something unknown in the obscure light, for example, of the moon. He will see various forms, shapes, objects, etc., which appear and then disappear, because they are not compatible with further experiences resulting from his movements, probings, etc. So there is a continual process of "trial and error" in which what is shown to be false is continually being set aside, while new structures are continually being put forth for "criticism". Eventually there develops in this way a perception which stands up to further movements, probings, etc., in the sense that its predictive implications are actually borne out in such experiences. (Of course, even this is always tentative, in the sense that it can be contradicted later.)

If we apply this reasoning to the subject of alien anomalies, we can say that humans first formed ideas about their behaviors through the lens of mystical, magical, or spiritual experience, and then later on through science fiction fantasy. These constitute "imagined characteristics" of these beings, instead of characteristics grounded in factual observations, such as would be the case with a being like a beaver, who we can observe over a steady period of time. Among these imagined characteristics of aliens was the idea that if aliens arrived on earth they would make their presence known in a global, unmistakable, and irrevocable manner, that they would be eager to communicate with us directly, and that they would be able to easily do so.

But now we might be seeing something far different. There certainly appear to be anomalies which seem "alien" to us. We have not yet really studied them and have mainly ignored them in fact. But from time to time, they remerge regardless of our interest. On top of that, they appear to be technological in nature and certainly do not strike the mind as having a natural or animal cause. Accordingly, their presentation leads some of us to naturally infer that these anomalies are the deliberate product of a "scientific" being similar in some undisclosed ways to ourselves. But that is about all we know of them at the present. It is also rather strange and curious to us that the makers of such technology would allow it to be seen and witnessed, while at the same time shielding their direct presence from us. Again, we do not understand "why" this is the case, but can only see that it appears to be so.

Those who are caught in the trap of science fiction fantasy conditioning, however, cannot stop their minds from continuing to speculate on their origins, convincing themselves that they are gaining some form of knowledge when all they are doing is multiplying speculation. They, like Fermi, are drawn into a neurosis that never ceases and never becomes productive. Because they do not see these anomalies conform to their science fiction fantasy expectations, they become confused, anxious, and neurotic. They never stop to think and question, however, whether their assumptions might be the source of the problem.

While it is certainly acceptable to start thinking about aliens from the standpoint of science fiction sources, it makes no sense to allow those sources to limit the range of inquiry. For example, we can acknowledge our natural expectations and desire that aliens would immediately join our world and teach us all of their secrets, but it would be childish to then demand that the universe actually conform to those expectations if the evidence is showing otherwise. Thus, in articulating these expectations as logical starting points based on our assumptions, we must also acknowledge that they lack any empirical bases at all and are simply things that we would like to see. Thus, if alien anomalies do not behave in ways that are consistent with these expectations, we should not take that as a sign that there is nothing there to see; instead, we should allow the phenomena to speak for themselves, and realize that perhaps our *a priori* assumptions are in need of revision.

For these reasons, it appears irrational to waste any more human time pondering this so-called "paradox" as if it was some sort of objective phenomenon. In truth, there is no real paradox here at all. Instead, we simply have a good, old fashion clash of paradigms. The unease and dissonance that we experience when we contemplate this strange situation is simply a signal that our childish paradigms about the nature of extraterrestrial reality are simply wrong. In order to begin the process of seeing things "correctly" (whatever that might really look like), all we need to do is understand the nature of the limitations of our previous paradigms, release our strong attachment to them, and understand that we have a choice concerning which paradigms we want to rely on.

3.3.3.2 The "universal scientific community" hypothesis

The SETI paradigm also rests on an assumption that the present-day human understanding of math and science represents an approach to fundamental reality which is complete in its approach, instead of simply reflecting back to the observer a limited slice of a much greater and indecipherable whole. Put differently, we can say that a key aspect of the SETI paradigm rests on an assumption that there is a common identity of interests, perspectives, and understandings, between human minds and alien minds in any situation in which human to alien communication could occur. We might refer to this as the "universal scientific community" hypothesis, which claims that all scientifically minded alien beings throughout the universe would share an understanding of math and science that is similar enough to what has been developed by humans in the late 20th century so as to make instantaneous and mutually significant communication possible.

It is true that we have no *a priori* grounds to conclude that simian lifeforms are unique to planet earth, or that the processes by which technology and evolution have unfolded here on earth are not universal in some sense; moreover, we have no reason to believe that our understanding and expression of mathematics and science is completely idiopathic to our species. At the same time, do we really know just *how* universal the processes with which we are terrestrially acquainted actually are? How many times have scientists convinced themselves that their understanding of the universe was true and complete only to be later shown that it was false? Why would humans in the early part of the 21st century think that their understanding has reached any sort of endpoint? And on this particular question, in the absence of any real alien examples to compare ourselves with, how could we be so sure of the completeness of our knowledge?

On an abstract level it seems plausible that other alien species in the universe *could have* evolved in ways very similar to us and therefore could practice math and science in the same ways that we do, and perceive and experience an external reality in a way that is mutually relatable to us. At the same time, we don't have any idea how intelligent alien life (if it really exists) is actually distributed throughout the universe, or what it actually looks like.[25] The assumption that aliens would be like us is an inference we draw from our local circumstances with absolutely no idea as to its universal applicability. As a result, it seems somewhat problematic to make reliable predictions about whether or not aliens like us are actually within a range that would make it possible for us to directly communicate with them within a realistic temporal horizon (even if that range is through deep space radio signals at its outer limit). We must admit that this is only an assumption, or hope of communication, and nothing more than that.

Think back to the example of the beaver for a moment. In that case, we do not expect to be able to sit down with a beaver and have a conversation with him or her about the methods they use, the choices they make, and so forth. We expect that we must gather that information on our own by observing the doings of the beaver and making deductions and inferences about what we think is going on. But we naturally have different expectation when it comes to extraterrestrial aliens for obvious reasons. We do expect that such a direct conversation is possible because we assume that their higher knowledge of physics also implies that they would be able to have the power of speech and communication in the same way that we exercise these facilities in and among our own species and that they would understand science in the same type of way. But is this anything more than a science fiction fantasy? Is it based on any real details or precise ways of thought?

---

[25] If aliens exist, there will be an actual, non-imaginative distribution of their presence which has been determined by the operation of both natural forces (such as exoplanetary development and evolution), as well as the spontaneous, intelligent choices of aliens, that have absolutely nothing to do with us. In the absence of any direct evidence of this actual distribution, any distribution we might think of is wholly speculative in nature (such as, for instance, the so-called Drake "equation" which is nothing more than a fact-free act of the imagination).

3.3.3.2.1 Examination of the assumptions behind the universal scientific community hypothesis

Let us now examine these types of assumptions in greater detail. We know that the Milky Way is approximately 100,000 light years across. Let us assume that there are intelligent alien beings (call them "Aliens X", or "AX") as a distance 1,000 light years from earth. Let us further assume that, relative to the typical lifespan of AX, a 1,000 light year journey is the kind of undertaking that an average AX could make from its home planet and back 2 to 3 times before dying. In other words, a journey from the home planet of AX to earth is a long one, but doable for the AX species nevertheless, similar to the long-distance nautical voyages of Europeans in the 15th century. Let us also assume that human beings are not the only form of intelligent alien life within the 1,000 LY distance, and there are other intelligent alien species within that distance as well. If such other intelligent beings did actually exist within the range of AX, we would have no knowledge of them at all because we are terrestrially bound.

The possibility of AX making a 1,000 LY journey 2 to 3 times before it dies already seems like crazy science fiction to a 21st century human. But even that type of journey is really nothing when compared to the actually known size of the Milky Way, which has a radius 100 times larger than the journey we have just thought of. So how many beings with the capability of AX would live within the 1,000 LY radius? And of those beings, how many of them would have evolved along simian lines, or other biological lines, such that they would be immediately relatable to us as Cocconi and Morrison imply all aliens would be?

At its crux, this question involves two interrelated issues. One of them concerns the extent to which our scientific knowledge of our own evolution and development permits us to make scientifically reliable conclusions about the characteristics of unknown aliens on unknown planets. The second concerns the extent to which we can reliably say that the human experience of math and science is truly universal, as opposed to idiopathic in some undisclosed ways. For if there are real alien scientists within the range of our potential experience that come from a non-simian lineage, such as the insectoid, avian, or reptilian lines, they might have a completely different relationship to physical reality and science that is so bizarre as to be unimaginable to us.

Any conclusions we might draw about alien beings at this point in time arise exclusively from the experience and knowledge we have derived from the only other scientific beings in the universe we are aware of: ourselves. Since any conclusions we might make about ourselves are based solely on terrestrial sources of knowledge (which, incidentally, have only been recently developed relative to the depth of cosmological time itself) an examination of the sufficiency and completeness of such knowledge is a necessary initial consideration. Let us then first consider what we might be able to know about the universality of our own experiences with mathematics and science, since these fields are taken to be the "standard bearers" of objectivity among our species.

3.3.3.2.1.1 *Limitations on the human experience of "objective" truth*

In *Road to Reality: A Complete Guide to the Laws of the Universe* (2004), Roger Penrose argues that mathematical and geometric truth appear to manifest "an objective external standard that is not dependent upon our individual opinions or our individual culture." This is an articulation of the conventional assumption that makes the universal scientific community hypothesis possible as it assumes that the 21st century human perception of mathematical, geometric, and scientific truth is absolute and universal.[26]

One of the reasons why this assumption makes sense to us is because well-established mathematical experiences demonstrate that various geometrical and mathematical expressions sometimes contain, at their inception, deeper levels of truth that do not reveal themselves for quite some time. These truths do not appear to be the product of subjective, human creation, but appear to arise from an external source, thus creating an impression of objective manifestation. One example of this kind of thing is found in the distinction between Euclidean and hyperbolic space. For reasons that are not clear, Euclidean space is a form of space that human beings seem to naturally orientate towards. Some of the basic expressions of Euclidean space are described in Euclid's five

[26] We might refer to this as "Lord Kelvin's Trap" as discussed in **Section 3.1.**

postulates. The first postulate asserts the existence of a straight-line segment connecting any two points. The second asserts the ability to extend without limit the length of any line. The third asserts that a circle can be created by a line segment between any two points. The fourth postulate asserts the equality of all right angles. Euclid's "first four postulates," Penrose writes, "basically encapsulated our present-day notion of a (two-dimensional) metric space with complete homogeneity and isotropy, and infinite in extent."

Euclid's fifth postulate, also referred to as the "parallel postulate", states that if one takes two straight lines (a and b) and intersects a third line (c) across both of them in a way that makes the interior angles created by the transection of (c) across (a) and (b) *less than* right angles, the two lines will always intersect at some point if extended far enough. Euclid's postulates were introduced in 300 B.C. and seemed intuitively true upon their statement. Even someone with no understanding of Euclid's fifth postulate would likely think the idea was true, because it seems to be true to us naturally, as there is something about the way we relate to space that makes the fifth postulate seem right. However, attempts to formally prove the truth of this postulate, and demonstrate the existence of mathematical forms which could confirm its validity, were elusive.

In 1733, a geometer by the name of Girolamo Saccheri attempted to prove the fifth postulate by way of contradiction, meaning that he assumed the postulate was false, and then attempted to show that the results of this false assumption were absurd and contradictory. However, instead of being successful in his effort to prove the fifth postulate by negative contradiction, Saccheri "found" hyperbolic space. As described by Penrose,

> Saccheri failed in his attempt to find a contradiction. He was not able to obtain a proof of the fifth postulate. But in striving for it he, in effect, found something far greater: a new geometry, different from that of Euclid—the geometry that we now call . . . *hyperbolic geometry*. From the assumption that Euclid's fifth postulate was false, he derived, instead of an actual contradiction, a host of strange looking, barely believable, but interesting theorems. However, strange as these results appeared to be, none of them was actually a contradiction. As we now know, there was no chance that

> Saccheri would find a genuine contradiction in this way, for the reason that hyperbolic geometry *does* actually exist, in the mathematical sense that there is such a consistent structure.

Some of the prime features of these "strange looking, barely believable, but interesting theorems" include the fact that in hyperbolic space the angles of a triangle are always less than 180 degrees, while the shortfall between the angles of a hyperbolic triangle from 180 is always proportional to its area (something which never happens to triangles in Euclidean space). Similar types of things happen with other rigid bodies such as squares and rectangles. In short, any rigid plane object that relies upon the parallel postulate is a different kind of object in hyperbolic space than it is in Euclidean space. At the same time, there are nevertheless similarities between Euclidean and hyperbolic figures. As stated by Penrose,

> Hyperbolic geometry . . . turns out to have all the properties of Euclidean geometry apart from those which need the parallel postulate. We can construct triangles and other plane figures of different shapes and sizes, and we can move them around 'rigidly' (keeping their hyperbolic shapes and sizes from changing) with as much freedom as we can in Euclidean geometry, so that a natural notion of when two shapes are 'congruent' arises, just as in Euclidean geometry, where 'congruent' means 'can be moved around rigidly until they come into coincidence.'

In the case of hyperbolic geometry, the differences between it and Euclidean geometry were not known until at least two millennia after Euclidean geometry was first articulated when it was shown that Euclid's fifth postulated disclosed hyperbolic space if examined in a certain way. We could say that Euclid's Fifth was an actual gateway into new orders of knowledge. The way we discovered what Penrose refers to as the "consistent structure" of hyperbolic space, leads us to believe that its existence is independent from us and completely objective. When Euclid formed his fifth postulate, did he have any idea that hyperbolic space was lurking within it, not to be revealed for nearly two centuries? It seems unlikely to us. It was probably the case that centuries of mathematical thought were required before hyperbolic space could be revealed from the fifth postulate, giving Saccheri an epistemological vantage point that Euclid couldn't even dream of—a vantage point which was wholly alien to that of Euclid.

Given these circumstances, it seems difficult to say that Euclid's fifth postulate is entirely subjective, and it appears as if Euclid tapped into some source of knowledge that is independent of what any human being might think of it; otherwise, how could one explain this type of case? Yet at the same time, it also seems like Euclid's formulation of the fifth postulate was essentially a subjective exercise in description, even if it was based on a rigorous examination and analysis of the characteristics of space as he and others of his time perceived them (and how we still seem to perceive them to this day). It is hard to understand exactly what is going on here, but it would strain credulity to simply say that Euclid just made up the fifth postulate out of thin air, given the manner in which its examination led to the discovery of hyperbolic space. It certainly seems that Euclid was tapping into something "greater than the self", some sort of "objective" source of knowledge, when he did his work.

In a different kind of way, mathematical truth also expresses a sense of universality that appears to be self-disclosing over time. An example of this type of thing is the mathematical phenomenon known as the "Mandelbrot set". The Mandelbrot set is a structure based on a mathematical formula of particular simplicity that demonstrates an incredible level of detail and sophistication which reveals itself to the human mind as the formula is plotted out on a physical plane (you can google "Mandelbrot set" to get a physical representation of what we are talking about here). While a human being, Benoit Mandelbrot, was the one who "discovered" the set, he did not embed the complexity and detail in the set itself. It appears as if the complexity was "already there" before the discovery was made by a human. Penrose puts it like this. "Those designs were already 'in existence' since the beginning of time, in the potential timeless sense that they would necessarily be revealed precisely in the form that we perceive them today, no matter at what time or in what location some perceiving being might have chosen to examine them." In other words, the Mandelbrot set seems to be a universal expression whose content does not appear idiopathic in relation to human beings but is rather something objective which any scientifically minded alien in the universe (including us) would see. From this, conventionalists such as

Penrose go on to conclude that any being anywhere in the universe would perceive it in the same way.

More examples of this kind of thing could be generated from the fields of geometry and mathematics, instances where humans have the experience that they are discovering things that appear to have an independently "objective" existence, a life of their own. "Real" numbers are another example. Real numbers arose when mathematicians and geometers attempted to derive numerical measurements for the actual distances of rigid bodies. The values for the numbers which arose, such as $\sqrt{2}$, could not be reduced into the simple, discrete integer format which the human geometers and mathematicians themselves wanted to use. It was almost as if the values for these numbers presented themselves to humans from a seemingly independent source, imposing it against the will of the mathematicians. Penrose describes this mathematical experience in the following terms,

> In the development of mathematical ideas, one important initial driving force has always been to find mathematical structures that accurately mirror the behaviour of the physical world. But it is normally not possible to examine the physical world itself in such precise detail that appropriately clear-cut mathematical notions can be abstracted directly from it. Instead, progress is made because mathematical notions tend to have a 'momentum' of their own that appears to spring almost entirely from with the subject itself. Mathematical ideas develop, and various kinds of problems seem to arise naturally . . . It is as though Nature herself is guided by the same kind of criteria of consistency and elegance as those that guide human mathematical thought.

From the standpoint of a human being, the truths expressed by mathematics and science sometimes seem to come from another "place" because the postulates sometimes contain a range of undisclosed truths whose reality and consequences are not discovered by humans for a long, long time. While humans may discover the initial thread of a certain mathematical or geometrical truth, they don't always create the whole cloth all at once, as seen by an unfolding disclosure process which is distributed over time.[27]

---

[27] For a discussion of this same concept from a different angle, the interested reader can consult Professor Madden's discussion of the

But does this human experience, while certainly valid, provide a sufficient basis for us to conclude that all alien minds would in fact perceive these forms of knowledge in a same or similar way? Are people such as Penrose going too far when they claim that any perceiving being would see things this way, "no matter at what time or in what location some perceiving being might have chosen to examine them?" Does this experience prove that we have actually tapped into something which is universal in way that would be similarly recognized by any alien mind anywhere in the universe? While our experience does imply that this is a possibility, how can we really say that this experience proves it is a certainty? How far can we rely on this type of mathematical experience to draw any reliable scientific conclusions about the universality of such an experience with respect to real alien minds?

The problem with trying to answer this question is that regardless of whether humans are really interfacing with a universal source, all of our perceptions, thoughts, and expressions of what we experience takes place in the world of a human embodied consciousness which is the particular product of a limited set of historical conditions, factors, and accidents, which must be initially considered *local* in character until we have other local instances to compare them to and from which we can draw generalized conclusions. Today we understand from the discipline of human neuroscience that there are a vast array of local forces and conditions that shape our perceptions of the world around us. Without delving into the particulars, we can say that our study of human neuroscience shows that there are real structures in the brain that shape thought and perception; thus, we can no longer rationally believe that our thoughts and perceptions operate in an objective and unconditioned way.

Accordingly, what we know about our own cognitive process and perceptions naturally implies the existence of an "alien neuroscience" supporting alien minds as well. This necessarily implies that we must have a real understanding of alien neuroscience before we would be in a position say anything meaningful about alien minds; otherwise, we would be guessing how alien minds operate based on speculation about the assumed similarities between human minds and alien ones. In the absence

---

"hyperobject" in *Unidentified Flying Hyperobject: UFOs, Philosophy, and the End of the World* (2023).

of any real information about alien minds, anything we say on this point must have an irreducible element of the arbitrary within it. Yet, the majority of conventional scientists are completely blind to this fact and far too willing to posit universal features of consciousness based solely on abstractions made about human minds; thus, they repeat the same type of error that assumes human consciousness represents the objective "gold standard" in terms of a universal form of consciousness. *See* **Section 5.3.5** (discussing the work of E.O. Wilson).

For instance, we have no idea why we naturally gravitate towards Euclidean space as the prime basis for our experience; nor can we exclude the possibility that an alien consciousness could naturally gravitate towards hyperbolic space as the prime basis for its experience of reality, with Euclidean space being secondary (i.e., as type of space that is engaged in the mind or imagination, as opposed to the physical body). If an alien mind did operate naturally in a hyperbolic manner, as opposed to a Euclidean one, how could we reliably make any clear predictions about the similarities and differences between alien minds and our own? How could we know how an alien mind viewing its world through a hyperbolic lens—as opposed to the Euclidean one we use—would perceive the world, or relate to mathematical or geometrical truths as we see them, since we do not have a direct experience of hyperbolic space ourselves? How might such an alien being perceive our expressions of what we see? Would a hyperbolic being even see a Mandelbrot set as we see it? Would it see the same real numbers in the same way we do? Or would it have a completely different type of numerical relation to reality in which things which seem intractable to us, seem natural to it, or vice versa? We are limited beings seeing a reality in a limited way and the things that we think are universal or clearly understood could have a far different meaning to a hyperbolic alien mind.

Unless we can articulate an adequate basis, for instance, to explain how a real and embodied hyperbolic alien mind would perceive what we Euclideans see on a basis of complexity and detail that rivals that of human neuroscience, it would seem difficult to have much confidence in our ability to know such a thing. Lacking direct knowledge about the degree of actual universality or similarity between these two modes, how far they would extend, or how they would affect the intimate and personal characteristics of such alien minds in relation to our own, it would be fundamentally speculative to make concrete predictions about the characteristics of hypothetical alien minds based solely upon our terrestrial experience and knowledge filtered through the lens of Euclidean space.

> The existence of various different models of hyperbolic geometry, expressed in terms of Euclidean space, serves to emphasize the fact that these are indeed merely "Euclidean models" of hyperbolic geometry and are not to be taken as telling us what hyperbolic geometry actually *is*. Hyperbolic geometry has its own "Platonic existence", just as does Euclidean geometry. No one of the models is to be taken as the "correct" picturing of hyperbolic geometry at the expense of the others. The representations of it that we have been considering are very valuable as aids to our understanding, but only because the Euclidean framework is the one which we are more used to. For a sentient creature brought up with a direct experience of hyperbolic (rather than Euclidean) geometry, a model of Euclidean geometry in hyperbolic terms might seem the more natural way around.

As Penrose himself notes in this passage, our experience as human beings gives us one particular vantage point on the world—that of Euclidean space, but we also know that hyperbolic space is real as well. Given our lack of knowledge of any sentient beings outside of this planet, can we confidentially state that our vantage point is universal? Do we really know today how different a world would be for a sentient being orientated primarily to hyperbolic space, as opposed to Euclidean? Can we even imagine such a thing as a real possibility within the vastness of the universe and the operation of cosmological time? Can we reliably describe the actual extent to which our view on reality is idiopathic or truly universal? Without clear answers to questions of this nature, any conclusions we might draw about alien minds could only be speculative or provisional in nature.

3.3.3.2.1.2 *Epistemological Limitations on our evolutionary knowledge*

In prior eras, it was assumed that human beings naturally possessed the "gold standard" of direct conscious perception and thought, and that our ways of thinking and being were modeled upon a universal standard handed down by a Divine Creator. While today we know this idea simply isn't true, its legacy lives on. Creation itself is certainly divine, mysterious, and appears fundamentally unknowable in all of its particulars within the parameters of human consciousness; moreover, a belief in a creator, or group of creators, that put this human reality together and maintain it somehow is far from foolish. At the same time, we can no longer ascribe any credibility to the position that everything about the way we experience and process the world represents universal characteristics as opposed to merely local ones. Put differently, today we can only say that the whole of human consciousness is simply one minor variation within the field of "consciousness itself". We can no longer rationally represent to ourselves that we deploy the most universal form of consciousness, it is a false conceit to claim that we do.

It is probably the case that features of our local species are the result of processes that possess some degree of universality; this, at least, is a fair inference based on our experience of things to date. Just as it would be foolish to think that the 21st century human experience is totally universal, it would also be foolish to say it was universally idiopathic. Both kinds of extremes would be unwarranted. Clearly, these higher order types of activities, such as the arts of manufacture and scientific thought, are tapping into something that is not merely local, and which at times has predictive power and apparent universal application. At the same time, we lack the sort of precise details (i.e. "real numbers") that would lead to real knowledge about the extent of the objective features of these modes, when compared to real and actual forms of alien consciousness. Accordingly, when we rely on our local frames of reference to make universal declarations about unknown alien minds, we must be exceedingly cautious as any such declarations can only be provisional in nature. In this respect, the SETI paradigm is incautious and sloppy, taking certain assumptions as articles of faith and jumping to unwarranted conclusions based on emotionalism, not Reason.

Complicating matters even further is the fact that we lack a precise accounting of how we actually and really got from a being such as Concestor 13 (or even a being more primitive than that) to Euclid and his fifth postulate. As Kuhn himself says, the first firm consensus in mathematics was "prehistoric", meaning that we really have no idea how this train of thought actually started in the past as we have no conscious knowledge of the specifics of this formation process within ourselves. Unless we are content to engage in gross fantasy speculation with respect to alien minds, we must possess details of this nature in order to assess the real similarities and differences between alien and human minds, as we are relying on the observable features of our own minds to draw conclusions about alien ones. Put differently, we certainly know that there was a path that got us from non-scientific creatures to scientific ones, but we don't know the specific, step-by-step details as to how beings go from shrew-like creatures incapable of any scientific thinking, to beings who are able to formulate rarefied statements like the fifth postulate. We do know a great deal from the point of Euclid forward due to the insights we may gain from studying the written record of thought. But events and relationships which occur prior to the written record remain a critical, yet undetermined and mysterious, piece of the puzzle.

We have very little "real knowledge" about how non-scientific beings become scientific ones, even though it appears to be the case that such a process does occur and gave rise to us. In other words, we no longer believe that scientific consciousness arose in a magical flash that we cannot study or understand, even though our understanding of the actual dynamics of that "flash" are hazy and inconclusive. Present evolutionary theory cannot provide us the material details and a coherent explanation about how we ourselves went from "no scientific thought" to "Euclidean or Pythagorean scientific thought". One would think that our understanding of our own evolution would be one area in which our knowledge was at its zenith, and that we should at least be able to describe to ourselves how we got to where we are today. After all, we have more direct knowledge about our own evolution and development than any other species on earth. But there are pervasive, material discontinuities in our understanding of our own development and evolution which call its completeness into question, and which should make us very leery and suspicious of drawing any conclusions with a respect to an analysis of unknown alien consciousness based solely on what we know about ourselves.

The most we know from conventional sources of knowledge is that there was an evolutionary "moment"[28] about 50,000 years ago

[28] The term "moment" is not meant to imply that the changed happened "all at once", but simply that there was some bounded time pressure that

which scientists call "The Great Leap Forward" when human culture exploded. Anatomically, no changes can be detected in *homo sapiens* from before and after this mythological event. But the archaeological record indicates that around this time everything changed for humans. Dawkins describes this moment as follows,

> Earlier than the Great Leap Forward, man-made artefacts had hardly changed for a million years. The ones that survive for us are almost entirely stone tools and weapons, quite crudely shaped. Doubtless wood (or, in Asia, bamboo) was a more frequently worked material, but wooden relics don't easily survive. As far as we can tell, there were no paintings, no carvings, no figurines, no grave goods, no ornamentation. After the Leap, all of these things suddenly appear in the archaeological record, together with musical instruments such as bone flutes, and it wasn't long before stunning creations like the Lascaux Cave murals were created by Cro-Magnon people. A disinterested observer taking the long view from another planet, might see our modern culture, with its computers, supersonic planes and space exploration, as an afterthought to the Great Leap Forward. On the very long geological timescale, all of our modern achievements, from the Sistine Chapel to Special Relativity, from the *Goldberg Variations* to the Goldbach Conjecture, could be seen as almost contemporaneous with the Venus of Willendorf and the Lascaux Caves, all part of the same cultural revolution, all part of the blooming cultural upsurge that succeeded the long Lower Paleolithic stagnation.

Dawkins reasonably speculates that perhaps this Great Leap Forward coincided with a "new software technique; maybe a new trick of grammar, such as the conditional clause, which, at a stroke, would have enabled 'what if' imagination to flower." Or perhaps, Dawkins suggests, "some forgotten genius realized the possibility of using words referentially as tokens of things that were not

---

caused a discrete change. How long that "moment" may have actually lasted is impossible to know because we were not there to witness it, and it might have occurred in a geological timescale that is simply not accessible to a human being who does not know how to directly experience time in that manner (as we have no direct experience of the real meaning of either cosmological or geological "moments").

immediately present."[29] These are certainly plausible suggestions, but as to what actually happened, and how it happened, that knowledge is shrouded from our view at present. In the gap that exists between the dead record of evolution and the active life process that actually caused evolution, a high degree of indeterminacy arises which shows itself repeatedly.

> The sweep of geological time is normally so far beyond the ken of poets and archaeologists it is daunting. But geological time is large not only in comparison to the familiar timescales of human life and human history. It is large on the timescale of evolution itself . . . If we measure evolutionary rates over a short time, and then extrapolate say, to a million years, the potential amount of evolutionary change turns out to be hugely greater than the actual amount . . . Darwinian selection, if we impose it artificially as hard as we are able, can drive evolutionary change at a rate far faster than we ever see in nature. To see this, we cash in on the lucky fact that our forebearers, whether they fully understood what they were doing or not, have for centuries been selectively breeding domestic animals and plants. In all cases these spectacular evolutionary changes have been achieved in no more than a few centuries or, at most, millennia: far faster than even the fastest evolutionary changes that we can measure in the fossil record . . . The waves and eddies of natural evolution are usually too slow for us to see in our little lifetimes . . . [however], the selection pressures actually to be found in nature, even if they don't always pull in the same direction, are orders of magnitude stronger than anything dreamed of by the most optimistic founders of the neo-Darwinian revival. And this again underlines the point: why doesn't evolution go much faster than it does?

In the evolutionary record we see a series of discontinuities and infer that a continuous and transitional process must have brought about those forms based on the perceived relationships between those forms. "People and chimpanzees are certainly linked via a continuous chain of intermediates and a shared ancestor," Dawkins writes, "but the intermediates are extinct: what remains is

---

[29] Referring back to the arguments in **Sections 2.16** through **2.18,** we note that Dawkins describes the important role language suggestions play on the field of human thought and real subsequent progress.

a discontinuous distribution." But we are unable to directly observe the process of species transformation itself. Instead, we rationally infer that there are array of chance contingencies which come into alignment, and that there must be a process by which those contingencies coalesce and become determinate, resolving into discrete forms. We draw this inference because we ourselves, as bystanders to this process, are able to both perceive separate species and to perceive the aftereffects of a process by which species transform and mutate.

> No evolutionist thinks that modern species change into other modern species. Cats don't turn into dogs or vice versa. Rather, cats and dogs have evolved from a common ancestor, who lived tens of millions of years ago. If only all the intermediates were still alive, attempting to separate cats from dogs would be a doomed enterprise . . . Far from being a question of ideal essences, separating cats from dogs turns out to be possible only because of the lucky . . . fact that the intermediates happen to be dead. Plato might see the irony in the fact that it is actually an imperfection—the sporadic ill-fortune of death—that makes the separation of any one species from another possible. This of course applies to the separation of human beings from our nearest relatives—and, indeed, from our more distant relatives too. In a world of perfect and complete information, fossil information as well as recent, discrete names for animals would become impossible. Instead of discrete names we would need sliding scales, just as the words hot, warm, cool and cold are better replaced by a sliding scale such as Celsius or Fahrenheit.

Put differently, due to our shortened lifetimes, and the relatively brief amount of time that we have spent studying the evolutionary history of life on this planet, we do not yet have a direct understanding of what "real" geological time actually looks like—instead, all of our knowledge of geological time is of an "imaginative" or inferential character, and therefore not capable of providing a precise readout of the process that actually occurred in the various evolutionary moments in time. This is most likely than not a function of the fact that our perceptions and understandings form within the timeframes of human lifetimes, which are far shorter than the cycles of time that the earth creates for life. In this regard, an analogy can be drawn between this limitation on actual evolutionary knowledge and some of the problems faced in the articulation of actual sub-quantum processes in the past. In *Causality and Chance* (1957), Bohm discusses one of these problems facing physicists in 1957 in the following terms,

> If there is some sub-quantum mechanical level, then . . . processes with very high energy and very high frequency may be faster than the processes taking place in the lower level. In such cases the details of the lower level would become significant, and the current formulation of the quantum theory would break down.
>
> For example, in our point of view, the "creation" of a particle, such as a meson, is conceived as a well-defined process taking place in the sub-quantum mechanical level. In this process, the field energy is concentrated in a certain region of space in discrete amounts, while the "destruction" of the particle is just the reverse process, in which the energy disperses and takes another form. In the quantum domain, the precise details of this process are not significant, and can therefore be ignored. This is in fact what is done in the current quantum theory which discusses the "creation" and "destruction" of particles as merely a kind of "popping" in and out of being in a way that is not supposed to be subject to further description. With very fast high energy processes, however, the results may well depend on these details, and if this should be the case, the current quantum theory would not be adequate for the treatment of such processes.

This example is the obverse of the limitation in evolutionary understanding but it speaks to the same basic underlying issue. When the quantum physicist cannot give a more precise description of the process by which particles "pop" in and out of existence, Bohm says that the description is inadequate because it fails to account for another level of a real process that must be occurring. Bohm's point is that there is some actual process occurring in the gaps before things pop in and out, and that process should, in theory, be measurable and understandable if we can see its fruits. Thus, when the quantum physicist says that the particles pop in and out, and is incapable of going deeper to describe how the pop starts and how it finishes, he is working with an incomplete paradigm. At the same time, the paradigm will still function properly so long as it is applied within limits that are appropriate to its techniques of knowledge. Indeed, one of the strange things about scientific paradigms of this nature is that they work even though they are always "wrong" or incomplete in some undisclosed ways.

A similar limitation applies to evolution, but in the opposite way. Here, the timescales are far too large, and we haven't been around long enough to see and study how species actually evolve in particular moments, when the time pressures bring things to a head and a new species "pops" out. We have not been able to gather the whole of time needed to bring those forms to fruition (think, for instance, of the use of time lapse photography to observe processes that are drawn out over time. Would it even be possible for a human to observe an evolutionary "occurrence" which took 100 million years to happen to a degree of detail that would give the observer complete knowledge concerning how that process actually happens? Yet, without being able to grasp the whole of this creation in one "moment", can we even really understand it? Would such an event ever be observable for a creature such as ourselves whose members struggle to live for 100 years at a time?).

Thus, while our evolutionary knowledge tells us certain things, it is also incomplete and cannot give us enough information to make any sufficiently reliable extrapolations on the question of how non-scientific beings transform into scientific ones. While the evolutionary record establishes beyond all doubt that this does happen, it does not establish precisely *how* it happens, that it has to happen as it did, or that it could not have turned out far differently. We lack a perspective of sufficient depth of knowledge to articulate a law of necessity here; all we know is that contingencies existed and were "resolved" in favor of particular trajectories, but we don't know the precise mechanisms of the step-by-step resolving process itself. We see the aftereffects of operations whose existence is implied, but not directly given to us, in a way that is similar to the 1957 quantum physicist who could see the particles popping, but couldn't describe how they actually got to popping. Our lack of knowledge when it comes to the subject matter of non-scientific animal minds evolving into scientific human ones indicates a cautious and modest approach regarding any predictions about the universality of human scientific experience in relation to unknown alien minds.

In this connection, it is perhaps also worth noting that the sudden explosion of human consciousness and culture is not the only anomalous occurrence contained in the evolutionary record. "If modern zoology admits of anything approaching a full-blown origin myth," Dawkins tells us, "it is the Cambrian explosion. The Cambrian is the first period of the Phanerozoic Eon, the last 540 million years, during which animal and plant life as we know it suddenly became manifest in fossils. Before the Cambrian, fossils were either tiny traces or enigmatic mysteries. From the Cambrian onwards, there has been a clamorous menagerie of multicellular life, more or less plausibly presaging our own. It is the suddenness with which multicellular fossils appear at the base of the Cambrian that prompts the metaphor of explosion." To be clear, Dawkins himself is skeptical of the idea that some mysterious, overnight explosion of evolutionary diversity actually occurred, and suggests that what appears as a sudden explosion is an illusion that exists for us because we are unable to realistically account for the changes that actually occurred in light of our lack of a clear and continuous scientific record which would show actual patterns of species popping in real geological time, as opposed to the imaginative geological time which we conceive of.[30] As stated by Dawkins,

> In the multidimensional landscape of all possible animals, living creatures are islands of viability separated from other islands by gigantic oceans of grotesque deformity. Starting from any one island, you can evolve away from it one step at a time, here inching out a leg, there shaving the tip of a horn, or darkening a feather. Evolution is a trajectory through multidimensional space, in which every step of the way has to represent a body capable of surviving and reproducing about as well as the parental type reached by the preceding step of the trajectory. Given enough time, a sufficiently long trajectory leads from a viable starting point to a viable destination so

---

[30] As Dawkins states, "It is all too easy, with hindsight, to think that because we recognize two ancient fossils as belonging to different modern phyla, those two fossils must have been as different from each other as modern representatives of the phyla are. It is too easy to forget that the modern representatives have had half a billion years in which to diverge. There is no good reason to believe that a Cambrian taxonomist, blessedly free of 500 million years' worth of zoological hindsight, would have placed the two separate fossils in separate phyla. He might have placed them only in separate orders, notwithstanding the then unknowable fact that their descendants were destined eventually to diverge so far as to warrant separate phylum status."

> remote that we recognize it as a different phylum, say, molluscs. And a different step-by-step trajectory from the same starting point can lead, through continuously viable intermediates, to another viable destination, which we recognize as yet another phylum, say annelids. Something like this must have happened for each of the forks leading to each pair of animal phyla from their respective concestors.
>
> The point we are leading up to is this. A random change of sufficient magnitude to initiate a new phylum at one fell swoop will be so large, in hundreds of dimensions simultaneously, that it would have to be preposterously lucky to land on another island of viability. Almost inevitably, a megamutation of that magnitude will land in the middle of an ocean of inviability: probably unrecognizable as an animal at all.
>
> Creationists foolishly liken Darwinian natural selection to a hurricane blowing through a junkyard and having the luck to assemble a Boeing 747. They are wrong, of course, for they completely miss the gradual, cumulative nature of natural selection. But the junkyard metaphor is certainly apt to the hypothetical overnight invention of a new phylum. An evolutionary step of the same magnitude as, say, the overnight transition from earthworm to snail, really would have to be as lucky as the hurricane in the junkyard.

Because we lack direct knowledge of the actual process of continuity that we rationally infer must be at play, the phenomenological perspective we presently have is peppered with mysterious discontinuities in a way that is similar Heisenberg's uncertainty principle (i.e., we can know some things we want to know, but not everything all at once). At some other level which we no longer have direct access to there must have been an actual set of non-mysterious operations in the active world because we can see the end products of a species' form. We know, in other words, that something real did form, and that there must have been a real process at some point that brought it into being; but as to the "particulars" of that real process, we are ignorant at the present time.

For the purposes of the present discussion, the point is simply this: if we lack direct knowledge of critical aspects of this time-dependent discrete, continuous process that must have given rise to our own terrestrial consciousness, then we must be very cautious when relying on such shaky foundations to convince ourselves that we would actually know anything real about alien minds, and wouldn't simply be making wild guesses. There are simply too many unknowns at this point in time (but maybe if our present state of knowledge was more developed, things would be different). Of course, there is nothing wrong with making wild guesses if one has no other rational option; the problem arises, however, when one is unaware that one is making wild guesses and delusionally believes that one is articulating the truth of things in a reliable way. When people assume that human consciousness and perception are completely universal in nature, they are making this kind of mistake. We simply can't assume that human consciousness is at the center of consciousness itself instead of simply being one local form of it. We also cannot be certain as to which aspects of our own consciousness are idiopathic or universal in nature (meaning that any other alien mind would see and understand things in roughly the same way as we do).

In this connection, Dawkins cites to a question posed by another biologist, Stuart Kauffman, on the question of whether or not we would see the same results we have today if we went back in time to earlier points in evolution and played the tape forward again.

> One way to underline our current ignorance is to ask, if evolution were to recur from the Precambrian when early eukaryotic cells had already been formed, what organisms in one or two billion years might be like. And, if the experiment were repeated myriad of times what properties of organisms would arise repeatedly, what properties would be so rare, which properties were easy for evolution to happen upon, which were hard? A central failure of our current thinking about evolution is that it has not led us to pose such questions, although the answers might in fact yield deep insight into the unexpected character of organisms.

Applying this statement to the "science fiction question of what life on other planets might be like," Dawkins notes, for instance, that the effects of gravity on other planets would have a huge and unpredictable impact on the unfolding of evolution. "On a large planet," Dawkins notes, "gravity would impose a whole new set of selection pressures."[31] While we could speculate on how those pressures would affect evolution, we need not do so in order to get the broader point as it relates to our knowledge of an alien consciousness; for if we can't conclusively determine how humans ended up with scientific consciousness on planet earth, it would be arbitrary to believe that our speculations about alien minds on alien planets we know absolutely nothing about are bound to be true. There are simply far too many unknowns at the present time in our state of knowledge for us to say our terrestrial sources of knowledge are complete and thorough enough to allows us to make reliable predictions about alien minds today.

If species simply provide us with a series of discontinuous examples of what evolution can do (without giving us direct knowledge as to how it is actually done), we understand that even within the terrestrial profile presented by earth there are a number of jumping off points for potential forms of scientific consciousness that would be truly alien to us. Just imagine, for instance, what kind of scientific consciousness would develop if a duckbill platypus one day produced its own version of a Euclid or Pythagoras. Platypuses have over 40,000 electrical sensors distributed in longitudinal strips across their bills, as well as over 60,000 mechanical sensors. These sensors provide the platypus with the electro-spatial equivalent of human binocular vision, allowing them to generate an electrical field in space to pinpoint the direction of their prey by direct detection just as we might do with our eyes. Dawkins likens this natural technology to a sophisticated "reconnaissance device, an AWACS organ". The brains of platypuses are wired to take the inputs from these fields and "calculate" the location of prey in a manner similar to the computer networks in human radar systems. Since a platypus is even more developed than Concestor 13, no one can say that such being could never attain to scientific consciousness within the breadth and depth of cosmological time (even if not within terrestrial geological time). One can only imagine what a scientific consciousness in such a creature would truly be like, or what technologies a Pythagorean duckbill platypus might concoct.

---

[31] This could also be true of the pressures exerted by time on an alien planet. Recent findings from the James Webb Space Telescope have revealed that our understanding of time the further we go away from planet earth is incomplete in ways that we do not fully comprehend.

Or, even stranger, imagine what a Euclidean or Pythagorean version of a star-nosed mole might do. The star-nosed mole (which is also at least as well developed as Concestor 13) has a highly sensitive sense organ in the middle of its "face" that "is more sensitive than any other area of skin anywhere among mammals, not excluding the human hand." As a result, the star-nosed mole likely "sees" through the tactile function which we would associate with touch, meaning that the layers and levels of environmental density and complexity that we experience through the eye, are likely experienced in the star-nosed mole through "touch". We tend to forget sometimes how optically dependent we are and how an optical interface with the world may not be the only possible way to interact with the universe. Pythagoras and Euclid were terrestrial masters of the optical realms, but in the face of the immeasurable and unknown galactic order of life we must be exceedingly cautious in concluding that the terrestrial forms which did manifest here on earth represent real limits on the possible. These are simply the limits and forms that did manifest on one particular planet, subject to a discrete set of forces that may or may not be universal in operation, and which we are only now beginning to understand. The forms we see on earth are certainly examples of what life can do, but they are not conclusive evidence of all that life can possibly do, even if it may seem that way to us upon our first impression of things. We have no idea how bizarre or unrelatable a star nosed mole version of a Pythagoras or Euclid would actually be.

Again, none of this is to say that either the platypus or star-nosed mole would actually become a Pythagoras or Euclid on planet earth in the future. However, both of these species are at least as developed as either a lungfish or Concestor 13, two species which we know did give rise to human scientific consciousness, so we cannot scientifically rule out the possibility of such a being on *a priori* grounds throughout the entire universe. Applying Nimoy's Dictum, one could say that such beings "are not outside the realm of possibility".[32] While we can say that such a form has not yet arisen on planet earth, we cannot say it would never do so in the future; nor could we say it cannot arise anywhere else within the depth of cosmological time itself.

---

[32] The phrase "not outside the realm of possibility," is colloquially referred to as "Nimoy's Dictum" in reference to the late actor Leonard Nimoy (1931-2015) who used this phrase repeatedly on the pseudo-scientific television show called *In Search Of . . .* (1977-1982).

3.3.3.2.1.3 *Limitations on our ability to know the real distribution of alien presence*

If we presume that aliens are real and do exist, then we must presume that there is a real distribution of their presence and that we stand in a particular relation to such a distribution instead of simply an imaginary one, and that it is the real relationship between us and them which determines if real contact will be made. But we have no basis to assume that every possible alien consciousness the universe may contain is within a range that we, or even our descendants, would have a real chance at making contact of some sort with it. It could be the case that, without us knowing it as we are barely able to leave planet earth, we live in a region of timespace where we actually are the only simian, Euclidean-orientated intelligent alien species, and the rest (if they really exist) have developed along evolutionary, technological, and cognitive/perceptual lines that are truly alien to us. We don't know, for instance, what a genuine and direct hyperbolic experience of space would really be like, or what Euclidean space would look like if viewed from the primary lens of hyperbolic space by an insectoid mathematician—and it could be the case that an actual alien of this nature is actually within our present-day range of experience. While something like this sounds insane, and it would be foolish to try to predict what an alien experience it would be to come face to face with such a being, nothing in our scientific ken says such a creature is outside the realm of possibility. To the people of Copernicus' day, the idea that the earth moved around the sun was probably no less fantastical that the possibility of an "insectoid Euclid" evolving on some other planet might seem to us. The strangeness and absurdity idea is simply not a good argument against it not being true.

3.3.3.2.1.4 *Why couldn't non-simian forms also develop scientific technology?*

In this connection, we cannot say that a simian is the only type of biological being in the entire universe that has the potential to develop advanced scientific technology. We know that our technological platforms are built upon the foundations represented by things such as the Pythagorean theorem, Euclid's postulates, levers, wedges, wheels, and metallurgy. The development of these technologies clearly involves some sort of paradigm-building, analysis, reckoning, and at least some sort of reasoning which we might call "proto-mathematical" and/or "proto-geometrical" in which a sense of abstract relations between objects are developed and explored that were latent within the environment but not brought into an explicit and actual format until the directed action of a human had been accomplished over a fixed span of time. This process initially starts with "found objects", evolving into the production and use of manufactured objects later on.

In the case of the wedge, a human obtains another object (such as a rock or piece of wood) places this object into the object she wishes to split, and applies force in a particular direction in such a way as to create a wedging effect which then splits the object into two. This behavior requires the use of some sort of paradigm which is drawn up in the mind of the *homo sapiens* and is not given by the environment itself. In the case of the lever, a human procures another object (let us say a piece of wood of some sort) and places it in relation to the object that she wants to move in such a way that the application of force to the object will now allow her to move it in a way that she could not do with unaided "brute" strength. The example of the wheel and axle present a more complicated arrangement because in that case there are additional levels of abstract reasoning and paradigm building which must occur. Not only must the human understand that the rounded use of the wheel gives her a certain new physical possibility of motion, but she must also think about designing an object which nature has not naturally placed in her pathway. However, as far as humans are concerned today, this type of behavior is not that difficult, and we would expect children to understand the idea based solely upon natural experience itself.

We can see in our own history a faint trajectory between these specific forms of paradigm development and execution and the advanced mathematics and science that we deploy today.[33] While these examples of "simple technology" seem very basic to us now, and we can reasonably infer that it is behavior of this nature that ultimately gave rise to the more complex forms of modern science, we don't know how precisely they came into being as we have no

---

[33] An example of this development is seen in the attraction called "Spaceship Earth" at Walt Disney's *Epcot Center*, which traces human development from its most nascent stages all the way through the dawn of the Space Age.

direct record or evidence as to the actual internal process in the minds of early humans as they developed these types of things since the time to observe the actual formation of these events has come and gone. And while we see examples of basic tool use among the present examples of our evolutionary ancestors—chimpanzees—we do not see the development of basic tool use beyond the most primitive forms. We have never, in fact, witnessed any type of species go from a state where it could not use levers, wedges, and wheels, to a state where it could design and use such objects like we do. Moreover, we certainly do not see chimpanzees and apes designing and implementing new technologies in an active sort of way. And we certainly do not see other simian versions of a Euclid and Pythagoras, except in science fiction movies such as *Planet of the Apes*. Thus, our real knowledge of the real origins of these techniques at the time they were first created is unclear and uncertain.

In addition to the uncertainties surrounding the specific development of our own scientific and technological capabilities, we do see within the range of other terrestrial creatures instances of behavior which may appear to the human mind to involve things like construction, adaptation, reckoning, and the use and transmission of geo-spatial information-from this, we can rationally imagine that these other terrestrial beings could provide a platform for the arts of manufacture, mathematics, and science in a far distant, perhaps unimaginable, future. We already discussed some aspects of this issue in the alien case of the beaver. While the activities of these creatures are certainly far less sophisticated than the type of activities we see in basic human technologies, geometrics, and mathematics, in some cases these creatures demonstrate a spontaneous ability to adapt the implementation of their paradigms to changing conditions of their environment that may appear similar in character (if not in degree of sophistication) to the type of capabilities humans display in the case of the lever and the wedge, and to some extent the wheel and axle. The difference is that these non-human creatures seemed to have plateaued in their capabilities and do not seem to continue to evolve in the same ways that we see ourselves doing. Because of this, we tend to assume that these creatures have reached the zenith of their evolutionary development. While it may appear that way within the parameters of present-day human time, we do not know for sure if any living creatures have necessarily exhausted their evolutionary potentials within the realm of geological or cosmological time, as they could undergo an evolutionary surge at a later date according to the laws of reason and the conventional knowledge we have developed from our study of evolution and cosmology.

Consider, for example, the male Bowerbirds of Australia and New Guinea. In *Animal Minds*, Dr. Griffin says these creatures construct "the most impressive structures built by any animals". The bowers are approximately 10 feet square, "supporting an avenue of woven sticks, with walls a foot high nearly joining in an arch over a floor, decorated with blue or green or natural or man-made objects. Perishable decorations, such as flowers, are replaced daily. Some bowers are left unpainted, but others are painted daily either blue, black, or green by the male, using a wad of bark as a paint brush and using crushed fruit, charcoal, or (nowadays near civilization) stolen blue laundry powder as paint." Scientists have observed that these bowers are used by the males to attract females, which will enter the bowers and sometimes allow themselves to be mated. Younger bower birds build less sophisticated bowers, and the decorative nature and structure of these bowers increases over the lifetime of an individual bower bird.

The bowers serve as a focus of immense activity for the birds. "An adult male," Griffin writes, "spends much time repairing his own bower, protecting it from raids by rivals, and attempting to steal ornaments or destroy rivals' bowers . . . Young males continually try to erect rudimentary bowers in the territory of an adult male but the latter patrols his property several times a day and wrecks them. When [a scientist] placed 100 pieces of numbered blue glass into these rudimentary bowers one night he found 76 of the pieces transferred to the bowers of dominant males by noon the next day." These bowers were apparently constructed for a "purpose" which was known among the birds, as evidenced by the fact that "[t]here was a statistically significant tendency for males with well-constructed and highly decorated bowers to mate with more females. The most attractive bowers were more symmetrical, included larger sticks more densely packed, and were decorated with more blue feathers, yellow leaves, and small shells." Indeed, it almost seems as if the birds understood the concept of beauty and were motivated by it.

Additionally, there appeared to be an active period of study for the younger male bower birds, by which they actually learned to construct bowers which were beautiful and might lead to success in attracting a mate.

> . . . earlier studies had shown . . . that many skills related to bower building and use have to be learned. Young males in green plumages spend about two years building rudimentary but increasingly complex bowers before acquiring blue adult plumage and building complete bowers. These 'practice bowers' are the joint efforts of several young males, which take turns placing and rearranging sticks, often clumsily and without success, occasionally with the cooperation of more skilled older males. The young males do not paint these bowers and are less discriminating than adult males in choice of color for bower decorations. Immature males spend much time watching the displays, mating and other bower activities of adults, and are displayed to by adult males.

The activity of these birds demonstrates a rudimentary ability to work with raw materials in the environment and shape those materials in such a way that they have a positive value that did not itself exist naturally in that environment, as well as a particular meaning within a network of communications among the birds. This display of knowledge constitutes a rudimentary form of manufacturing. Regardless of whether these birds are actually engaging in mental processes analogous to human thought, planning, or reckoning, when we look at their constructions we naturally see, without any explanations directly from the birds themselves, bowers which appear designed and well-ordered, and bowers which appear less so. This discrimination arises naturally in the minds of the scientists themselves, without a direct communication from the birds telling the scientists what to see, and appears to bear some correlation to what the birds are seeing amongst themselves. We see the birds responding to the same features in the bowers that we also see, as evidenced by the fact that scientists noted females are more attracted to what appears as "well-constructed" bowers. From a purely objective viewpoint, without resort to any speculations about what might be going on inside the "minds" of the birds themselves, we see an external manifestation of a pattern that has meaning to us in one way, meaning to them in another, and a common frame of reference among us both (i.e., "well-constructed") that appears to arise independently from either species. Clearly, in order to construct a well-ordered bower and appreciate a well-ordered bower, some form of knowledge must exist for these birds that was not naturally given to them in the environment itself. Just like the hurricane in Dawkin's junkyard did not construct a Boeing 747, so too did these bowers not simply "construct themselves" from the operations of the mechanistic environment *sua sponte*.

Of course, it would be extremely difficult, if not impossible, for us speculate as to whether the birds themselves perceive the structures to be "well-ordered" in the same way that we do, as this sort of knowledge would require us to be inside their minds. But they must see something positive they are responding to, and clearly a positive response is tied to the degree to which the bowers display a particular form of order that was not inherent in the environment itself in the absence of the directed work activity of the birds, whose knowledge we see reflected in the bowers which we perceive. This speaks to some sort of ability to go beyond the paradigms given by the natural environment and bring new paradigms into existence which constitute displays of active knowledge.[34] We could refer to these as acts of "creature creations." Such displays cannot be entirely explained by reference to mechanistic, autonomous, or involuntary instinct (i.e., "genes") because the displays are complex and respond to the changing demands of the environment over time—including changes that are completely alien to the birds, such as scientists intentionally placing alien objects in their midst in order to see how they might react.

Another example of this kind of knowledge capability is found in the world of the bees and the communications associated with the beehive. Humans and bees have been a part of each other's environmental milieux for at least 8,000 years, if not more. There is an 8,000 year-old cave painting in Arana Cave near Valencia, Spain, that depicts a human gathering honey from a beehive. Anecdotally, beekeepers have always noticed that bees would move over the surfaces of honeycombs in what appeared to be agitated "dances". It was also noted on many occasions that once a single foraging worker discovered a food source, many other bees from that hive would soon arrive at the same source in a

---

[34] We could apply this to how an alien mind might view products and expressions of our math and science. Surely, they would see something ordered, but can we reliably say that they would see order in the same we do? We know that the order makes sense to us in a certain way, but we cannot assume that to be the case with all alien minds.

manner that suggested knowledge regarding the location of the food source had somehow been communicated to other members of the hive.

However, it was not until the early 20th century that human scientists began to understand the nature of the communication mechanism that was actually in operation within the hive. This was due to the work of the Austrian zoologist, Karl von Frisch. Von Frisch began his work by establishing that bees could discriminate colors. He then studied the dances of bees by constructing beehives with glass windows to give him an entre of observation on their activities in the hives. Von Frisch also marked certain bees with dabs of paint, allowing him to identify individual bees so that he could track their movements. Because he took these specific measures, Von Frisch was able to visualize the otherwise invisible patterns of behavior and derive new knowledge about the nature of the bees' activities.[35]

Once Von Frisch brought these layers of invisible activity into human reality, he was able to observe that the bees returning with a sugar solution he had placed by their hive would engage in "round dances, moving in circles with alternative clockwise and counterclockwise directions" upon their return. Von Frisch also observed that bees returning with pollen (as opposed to the sugar solution) engaged in a different pattern of dance which he called the "waggle dance". Griffin describes the nature of these observations as follows,

> In a waggle dance the bee walks rapidly in a straight line while moving her abdomen back and forth laterally at about thirteen or fourteen times per second. Then at the end of this straight wagging run she circles back and repeats the straight part of the dance, followed by alternating clockwise and counterclockwise returns to the starting point of a series of straight wagging runs. Although the bee may move a short distance between successive cycles of the waggle dance, the basic pattern is relatively constant under given conditions.

---

[35] Drawing on points we made in **Sections 2.4** and **2.10**, we could say that Von Frisch's actions took the invisible world of the bees' activities and made them visible to human perception, which, as we noted in **Section 3.3.3.2.1.2**, is primarily optical in nature (these orders were not, of course "objectively invisible", as they were well known to the bees themselves this whole time).

> Von Frisch thus concluded that the two types of dances were somehow related to the sort of food being brought back to the colony, a reasonable interpretation of the observations he was able to make at the time.

Von Frisch conducted his initial research prior to WWII. After the war, he moved his laboratory to a new location in a more rural setting. By happenstance, he placed his feeding dishes a considerable distance from the beehive. He then discovered that bees who had to obtain sugar at more than 100 meters distance from the hive did waggle dances upon return as opposed to the round dances. The change in the dance patterns used by the bees implied that some kind of knowledge was being transmitted from the bees who had gleaned the knowledge from their travels to the bees who did not have this knowledge because they remained in the hive. Von Frisch ultimately concluded that waggle dances were being performed when bees were engaged in gathering materials from great distances from the hive. In other words, the waggle dancers were transmitting knowledge about the relative distance from the hive to the food source. This is self-evidently a display of knowledge demonstrating communication regarding measurement.

After decades of additional observation, Von Frisch learned that not only did the waggle dances communicate information about the distance of food, pollen sources, and other important information related to hive regulation, but that the dances also communicated directional information about the location of food or pollen sources as well. Griffin refers to this as "symbolic geometry," and describes it as follows,

> The most significant of von Frisch's discoveries was that the direction of the straight wagging run communicates to other bees the direction that the dancer has flown from the hive to the source of food. When the food is in the direction of the sun, the waggle dances are orientated approximately straight upward on the vertical surface of the honeycomb. If the food is located in the opposite direction from the sun, the dances point straight down, and if it is 90 degrees to the right of the sun, the dances point 90 degrees to the right of vertical. In other words, the orientation of the wagging run relative to straight up approximates the angle between the direction toward the food and toward the sun. When the sun is hidden

by clouds that cover only part of the sky, the honeybee visual system is able to achieve the same orientation by utilizing patterns of polarization in the blue sky that are correlated with the sun's position.

In other words, not only do bees possess at least some basic capability to communicate knowledge about distance, but they also possess the ability to communicate knowledge about crude and primitive angular measures as well. Additional displays of knowledge can also be witnessed in the activity bees in engage in when looking for replacement hives. When hives become overcrowded, older worker bees will cease looking for things like food, pollen, or water, and begin searching for cavities where a new hive might be established. They will often search over large areas, crawling into crevices in trees, rocks, or buildings, conducting what amounts to a site survey of places where new hives may be established. The new cavities "must be of roughly the right size and have only a small entrance near the bottom. It must be dry and free of ants or other insects," Griffin says.

Once these worker bees have concluded their survey, they return to the hive and "report back" to the hive members about the knowledge they have gained from their direct observations. They do this through the use of the same type of waggle dances that they previously used to report on the location and distance of food sources. When transmitting information about these new hive locations, the waggle dances indicate the location and distance of the potential hive site, and what appears to human scientists to be information about the potential hive site's "suitability". The features of a new location that make one site more suitable from another involve a variety of interrelated factors. After searching for new hive sites, dozens of scout bees may be involved in surveying several locations, when they report back, "the intensity of the dances is correlated with the quality of the cavity. Small, damp, or ant-infested cavities produce only a few feeble dances, whereas others that are dark, dry, and of suitable size lead to prolonged and vigorous waggle dances. Since different scouts have visited different cavities, a wide variety of locations are described by numerous dances that go on over the surface of the swarm."

Once the bees have generated and synthesized knowledge about the suitability of the various sites, and have transmitted this information through the hives, a collective decision-making process is initiated by hive as a whole. In the first few hours of the dances, a variety of locations are described; but over time, the number of described locations deceases. "[T]he cumulative effect of extensive dance communication," Griffin writes, "was a progressive reduction in the number of cavities described. And those that continued to be danced about were the best ones available." This communication process would continue over several days until nearly all the dances were about a single hive location. Griffin describes the culmination of this communication process as follows,

> After this had been going on for several hours, a different sort of behavior occurred, which Lindauer describes as a "buzzing run". The bees making these buzzing runs moved for fairly long distances over the swarm while emitting a buzzing sound. When this had been going on for some time the swarm took wing and flew fairly directly to the cavity that had been described by the concentration of enthusiastic dances over the past day or so.

As this study shows, bees possess a basic ability to generate knowledge about distances and angles ("measurement"), generate knowledge through basic investigative and observational processes, communicate forms of knowledge to other distinct members of their hive, and then engage in a collective decision making process in which knowledge is analyzed in at least a rudimentary fashion to arrive at a collective decision about where the hive should go, which is itself another form of knowledge that has itself been derived through some kind of indeterminate process involving all the separate members of the hive. Given the chain of events through which this process occurs, and the fact that the knowledge is derived from observations made by the bees of external conditions which change over time or are of an indeterminate nature, this activity cannot be reduced wholly to automatic or instinctual processes, which are only capable of responding to determinate stimuli in pre-determined ways. It also shows that these forms of knowledge, and the ability to generate paradigms and apply them, are not exclusive to the simian order or the human mind.[36]

---

36 In this connection, recent studies published in the journal *Science* suggest that avians and mammals have developed their forms of intelligence independently from one another. Gattoni and Tosches,

While the bees are able to generate and communicate geo-spatial knowledge within their own species, the honeyguide birds of Africa are actually able to communicate geo-spatial knowledge to species that are alien to their own order (i.e., humans). These birds locate beehives in the wild, and then guide human beings (and perhaps other animals, such as the ratel or "honey badger"[37]) to the locations of those hives. The process has been thoroughly documented by scientists, leaving no rational doubt that it is real. When the honeyguide wants to bring a human to a beehive, it attracts its attention "by a repetitive series of churring notes or it stays where it is and begins calling these notes and waits for the human to approach it more closely." It will then engage in a display for the human, involving a pattern of flight and tail feather movement. Once the human is drawn in closer, the honeyguide will then begin the "guiding" behavior of flying in the direction of the beehive from tree to tree, using churring sounds to attract the attention of the human, leading the human towards the beehive. Once the human reaches the location of the hive, the honeyguide becomes silent, waits for the human to take the honeycomb, and then feeds off the remains that have been left behind.

> Guiding may cover a duration of from a few seconds to half an hour, or possibly even an hour, and may involve a distance of from a few feet to over half a mile, and possibly, at times, even a mile . . . Guiding leads to the vicinity of a bees' nest, not to the exact spot . . . If a person does not follow a honey guide that has apparently come close to "lead" him, the bird may increase the tempo and excitement of its behavior as if to urge and entice, or it may give up easily and leave . . . . A would be "guiding" bird may sometimes follow a person for a very long distance (five miles is the maximum known to me) or for a very considerable period of time (half an hour is the maximum I know of) to attempt to get him to follow it.

Constrained roads to complex brains, Science, Vol. 387, Issue 6735 (February 13, 2025)

[37] The honey badger is also an amazing creature from the standpoint of an ability to adapt to alien situations. One can google honey badger on the internet and see videos of honey badgers opening locks on gates, using tools dropped in enclosures to escape, and a wide range of other spontaneous puzzle solving behavior.

Studies of the behavior involving the Boran peoples of Africa have left no doubt that the guiding behavior decreases the time that it takes for humans to find beehives and is effective. The honeyguides have been shown to have an accurate sense of direction that they are able to communicate to the humans by the guiding behavior. "The results clearly confirmed that the starting direction was indeed almost always correct within 20 to 30 degrees, and the mean direction of all initial flight directions was within less than one degree of the bearing of the bees' nests to which they eventually led the human cooperators."

It is a mystery as to precisely how these honeyguides form a paradigm linking the geometry of a spatial location to an information network that they can communicate to human beings. One anecdotal example from Dr. Griffin's book suggests that the way these honeyguides gather and organize this information may be through a mode that is alien to us.

> Captain Davison . . . had a group of his natives at one of the rest camps in the Reserve when a honeyguide came to them and chattered and went through all the motions of trying to get them to follow it. Davison refused to let any of his boys go, but got them all on a truck and drove off to the next camp some five miles away. The bird followed them all the way and then Davison told one of the natives to get an axe and follow the bird. The honey guide "led" this native to a bees' nest less than half a mile from the second camp, but which must have been at least 4 miles from where the bird first began calling to them.

One would naturally assume that before the honeyguide attempts to communicate the location of the hive to the human through guiding behavior it has surveyed the area, learned the location of the hive, and has committed that location to memory at some level. However, this anecdote (if true) implies something different may be at work because in this example the honeyguide seemed to know the location of a hive that was several miles away from the location where it first attempted to entice the humans into following it. Did the honeyguide already know the location of this second hive when it first approached Davison's men? Or, once the honeyguide was able to entice Davison's men, did it then somehow gain information about the location of another beehive without actually going out and directly observing it? Obviously, there would be no way to answer such a question in the absence of direct communication from the birds themselves, but the anecdote, if true, is intriguing.

The bower birds, bees, and honeyguides, serve as rare displays of non-human terrestrial, intelligence which perhaps show the potential for a development of a type of intelligence that could ripen into a form of primitive intelligence similar to that which allows a human to use a wedge or a lever.[38] They are anomalous and not characteristic of what appears to us to be the conventional features of non-human intelligence as it stands today, which we would generally term "instinctive" or "automatic", implying that such intelligence is incapable of responding to the ever-changing demands of external situations in the spontaneous, deliberative, and creative manner that we tend to associate exclusively with human thought. Of course, we cannot say for certain that these behaviors would one day ripen into the arts of manufacture and science, but if we consider these behaviors thoughtfully, can we deny such a possibility completely? For instance, these creature activities certainly speak to some ability to gather or "encode" and transmit knowledge regarding distance and geo-spatial relatedness, a basic ability to describe "where things are" to others. Couldn't this be seen as a crude and rudimentary expression of the type of "cognitive" process that underlies geometry and mathematics? We have been conditioned for millennia to believe that human intelligence possesses unique capabilities of reasoning, reckoning, measuring, and paradigm usage and development, which are exclusive to our species. But this could be an illusion. It could be the case, for example, that there are many non-human beings alive today who are displaying seeds of knowledge that could ripen into Pythagorean or Euclidean forms within the framework of geological or cosmological time, even though at present all we can see are the barest traces of the potentials.

In any event, we must concede that our range of knowledge is very limited compared to the scope of geological or cosmological time and the work that these forms of time can do. It could be the case, for instance, that it would take many generations of scientists to

---

[38] The wheel and axle may be slightly different case due to the particular way in which that technology is designed and used in the environment in an active, on-going way. But *Animal Minds*, does contain many examples of other non-human species that engage in active tool design and use more similar to the wheel and axle.

even begin to understand the possibilities of knowledge among life on earth because it would take that long to observe other non-human species actually changing in order to recognize the forms of knowledge they are capable of displaying. Consider how long it took, for instance, for Von Frisch to understand what bees had been up to right under the noses of humans for at least 8,000 years. Imagine what our understanding of life on earth would be like in another 8,000. There may be many forms of non-human knowledge which will remain hidden to us until we take the time to see them and we have no rational basis upon which to conclude that we have exhausted all sources of terrestrial knowledge at this point in time. We may have to wait until we can actually witness evolutionary changes and adaptations in real time at many different levels (including the genetic and molecular), before we can acquire knowledge at a scale and detail that is necessary for us to actually understand how life is changing and evolving on an actual scientific basis.

In spite of these limitations, if one releases the assumption that human thought is "at the center" of all forms of knowledge, then it is perhaps possible to see that at least a few non-human species display behaviors which naturally suggest that they "understand" how to formulate paradigms, deploy them in an active manner, and modify them in response to changing conditions.[39] The ordered display of these behaviors suggests that a specific content is somehow associated with these paradigms, including geo-spatial knowledge, group decision-making processes and communication, and manufacturing conduct. The bowers of the bower birds, for example, naturally appear to us as being "well ordered", and we do not require a direct communication from the bower birds to tell us that. We can see that the bower birds themselves have an analogous form of understanding because the more well-ordered the bower, the more attractive it is to females, implying that they notice the degree of order too. The bees appear to have some sense of measuring distance, reckoning angles, conducting examinations, engaging in group decision-making, and an ability to communicate knowledge generated from these activities among their own species. Again, we see this naturally from our own observations and do not need to communicate directly to the bees themselves to see that they know something. Honeyguides, by

---

[39] While these individual cases by no mean exhaust all the possibilities, they do demonstrate what kind of things are allowed in our world.

contrast, do directly communicate geo-spatial knowledge to us, and show us directly what they know, even though we are an alien species to them, occupy a position of superior understanding and clearly know a great many things that are far, far outside their imaginations.[40]

Could these behaviors be precursors to the development of thought patterns similar in complexity to those displayed by Pythagoras or Euclid, given enough time and the right evolutionary conditions? This suggestion carries an air of implausibility since we do not see insectoid or avian versions of Euclid or Pythagoras sharing the earth with us today. However, this does not mean we can reliably exclude this as a real possibility on *a priori* grounds, because in order to do so we would at least have to understand with reasonable precision how Euclidean and Pythagorean thinking came to be in human beings, so that we could then explain why these types of birds and bees could never arrive at the same state of knowledge. Furthermore, we must remember that the forms we see today cannot be said to represent the culmination or end stage of an evolutionary process within the scale and scope of geological or cosmological time. It is more accurately thought of as a snapshot of a particular moment in time. We don't know today, for instance, how things could change in 10,000, 1,000,000, 10,000,000, or 100,000,000 years—all of which are reasonable scales within the larger timeframes, even though they are practically outside of our own imaginations in a certain sense. We would likely need to observe these species for a period commensurate with that of geological or cosmological time to have some reliable knowledge regarding the evolutionary potentials at issue here.

3.3.3.2.1.5 *The "divine moment" error*

We cannot make universally reliable extrapolations about alien minds based solely on what we know about ourselves today because our understanding is too limited and incomplete. An insistence that what we see "now" represents all that evolution is capable of doing, leads one to repeat the same "divine moment

[40] For instance, if someone really believed that a honeyguide bird could form a meaningful understanding of what a computer is we would rightfully think them to be mad.

error"[41] that inhibited investigations of a Darwinian nature in the first place. We have no reason to assume that our presence as observers on earth has any geological significance whatsoever, or that what we see today is the end all and be all of what life has to offer on this world of ours.

Knowledge of evolution derives from two different orders: the "dead record" of the past, where we learn how life has likely played out on earth over time, and the "living record" of the present, where we see how life behaves today. Based on the interplay of these two orders, we develop an understanding of the probable trajectories of terrestrial life. However, real evolution is a time-dependent actual process which occurs in "moments" where different forces and factors coalesce, fall apart, die, or survive, involving a complex web of connections between the living organism and the environment in which it operates. Since the only moments of life we know are the ones we directly experience, the manner in which things really happened in other times is accessible to us only by inference and analogy, as opposed to direct evidence. Our knowledge about the past relies on an assumption that the hidden conditions which drive life today were similar enough in the geologic past to allow us to draw certain conclusions; while this may or may not be true, we haven't had enough time to witness species differentiation in real time to know if it is. This places a significant constraint on the completeness of our knowledge. While we can make reasonable inferences about the life processes in other times and places, because we lack a direct, realtime vantage point to see how those step-by-step processes actually played out, we don't know for sure why life took a turn in any one direction as opposed to any other—especially, when we go back further and further from the present.

In other words, at present, evolutionary biology does not tell us exactly *how* one species transformed into another, only *that one species has somehow transformed* into another, except for any

---

[41] The theory of evolution as articulated by Darwin (and others before him) was a counter-paradigm to the once dominant view that all life was created in a divine moment and that there would be no good reason to study its development since it was the exclusive domain of the supernatural being that created it in one "perfect moment" of divine time. Thus, we should not repeat the error that Darwin saw through by assuming that our brief presence on earth—and even briefer period of time engaged in the project of scientific thought—has given us exhaustive terrestrial knowledge.

species whose specific evolutionary arrival we have directly observed from a prior form. So, for instance, while evolutionary knowledge is conventionally depicted through the use of an image that shows the form of a lowly ape gradually evolving into a standing modern human, the image is misleading to the extent it implies that we know how "lower" forms evolved prospectively into "higher" forms. The image does accurately depict the bare facts that some later forms of life were necessarily preceded by earlier forms that shared similar structures, but as to the particulars of the actual process, that remains a somewhat of a mystery.

To emphasize this point, consider the following. Apes and chimpanzees are more similar to us than any other creature, and it appears self-evident that we evolved from them. But the characteristics that make us completely human are difficult to discern in them.[42] In 1999, the chimpanzee Ai was taught to identify numbers on printed cards and to select the card that corresponded to the number of objects shown to her. She was then taught to touch Arabic numbers on a computer screen in ascending numerical order. Once the numbers were touched in the correct order, she got food. The computer program would vary the location of the numbers from a sequence that she was familiar with. When this happened, Ai was observed to hesitate and make errors in an effort to get the correct sequence again. The circumstances of the experimental results did not appear random and would be very difficult to explain by reference to pre-programed instinct as the conditions of this experiment could not be foreseen by prior chimpanzees in Ai's evolutionary ancestry. The presence of error in the choices of the apes speaks strongly to the existence of a process of internal deliberation. The results led the researchers to conclude that "a three unit ordering task is supported in the chimpanzee, much as it is in humans, by planning, executing, and monitoring phases."

Clearly, apes have the capacity to understand numbers and work directly with our alien technology. Yet, for all the times that we have seen apes taught these skills, we have never seen the apes take this basic capability and then run with it and develop it beyond what their human companions and observers have taught them. We have not seen them turn to us and ask us to teach them how to do

---

[42] An excellent literary example of the "fully human" ape can be found in Will Self's novel, *Great Apes* (1998).

more of this kind of thing. We have not seen the rise, for instance of a Euclid or Pythagorean chimp (except in science fiction movies such as *Planet of the Apes*); nor have we *yet* seen chimpanzees on other places on earth start turning into humans.[43] While apes can be conditioned to learn skills, and even develop a spontaneous ability to use the skills in ways that humans did not directly teach them, we also do not see a great interest displayed by the apes in taking the knowledge that they have learned to use and pushing it beyond its present bounds. Lower order simians seemed plagued by an abiding apathy when it comes to the desire for knowledge that seems necessary to give rise to a non-human Pythagoras or Euclid.[44] Ironically, this apathy appears similar in nature to the apathy many conventional humans demonstrate with respect to alien anomalies today.

Dolphins, by contrast, do seem to spontaneously evidence an interest in learning things on their own. For instance, consider these sets of observations from Griffin's text, discussing the dolphins' capacity for the spontaneous imitation of behavior of other creatures not of their species.

> The behaviors imitated included the seal's swimming movements, sleeping posture, and comfort movements (self-grooming). The dolphin, like the seal, at times swam by sculling with its flippers while holding the tail stationary. The sleeping posture imitated was lying on one side of the surface of the water, extending the flippers, and trying to lift the flukes clear of the water. The comfort movement mimicked was vigorous rubbing of the belly with the under surface of one or both flippers . . . Additional imitative behaviors by this dolphin included the swimming characteristics and postures of turtles, skates, and penguins. The dolphin also attempted to remove algae from an underwater window with a seagull feather, in

---

[43] If humans of the future were to directly witness and observe chimpanzees transforming into evolutionary forms similar to *homo sapiens* it would constitute a substantial revolution in our evolutionary knowledge.
[44] As an aside, for the first time in history, biologists recently (i.e., in 2025) witnessed an orangutan using a medicinal leaf to patch up a facial wound. This example of spontaneous medical activity perhaps implies that there are a range of latent possibilities in the simian family that have yet to manifest. It could be the case that, if we were able to observe the orangutan over a geological timescale, we might see a growing range of complex, paradigm-driven behaviors develop.

> imitation of the activity of a human diver who regularly cleaned the window. The dolphin while "cleaning" the window, reportedly produced sounds resembling those from the demand valve of the diver's regulator and emitted a stream of bubbles in apparent imitation of the air expelled by the diver. Tayler and Saayman also observed a dolphin using a piece of broken tile to scrape seaweed from tank bottom, a behavior apparently derived from observing a diver scraping the bottom while cleaning the tank with a vacuum hose. The scraping behavior of this dolphin was then copied by a second dolphin.

In addition to this type of spontaneous imitative behavior, dolphins have also been observed to train themselves in tasks that are not natural to their species, as demonstrated from the following example,

> In one interesting episode illustrating self-training, one animal of the group had been taught to leap to a suspended ball, grasp it with its teeth, and pull it some distance through the water in order to raise an attached flag. This animal was subsequently removed from the show and a second animal in the group was trained in the same task, but learned to raise the flag by striking repeatedly at the ball with its snout rather than by pulling it. This second animal, a female, subsequently died and another female of the group immediately took over the performance, without training, and continued to strike at the ball with the snout. Later when this new female refused to participate in the show during a two-day period, a young male in the group immediately performed the behavior, but grasped and pulled the ball with his teeth, in the manner of the originally trained animal.

One could plausibly state that human scientific consciousness requires advanced and sophisticated versions of the apes' ability to count and reckon, and the dolphins' ability to spontaneously learn and respond to the changing conditions of the environment merged into one organism. Yet, neither of these species display the combination of these general characteristics in the way that we see in ourselves. At a certain level, it almost strangely appears as if we are hybrids of these two forms, a statement that makes absolutely no sense at all. No one in their right mind would say that humans evolved from dolphins, yet they naturally display characteristics that we know we have, and which other simians do not. Clearly something happened between ourselves and the lower simians which we do not understand.

The examination of these conventional sources of knowledge shows that we do not understand precisely how we became the scientific beings that we understand ourselves to be. We may know what we are at the present moment and have a general awareness of our own capabilities, but scientifically speaking, there are still a great many gaps and discontinuities in the field of our knowledge. But the modern, conventional scientific mind tends to forget this and often falls into Lord Kelvin's trap of thinking that we basically have everything figured out, when the truth is that we know very little at all.

3.4 Conclusion: the SETI paradigm is fundamentally arbitrary in nature and does not provide a reliable starting point for real knowledge about aliens

Returning then to the question of whether or not the SETI paradigm's assumption that aliens would be somewhat like us or see the world in a way similar enough to us to be instantly relatable, we can see that it fundamentally revolves around the possibilities of evolution on alien planets. Since we have no information or knowledge about evolution on alien planets, we are restricted to an examination of our own situation here on earth, making the best inferences that we can about what can happen elsewhere. In thinking about our own lineage of scientific knowledge, it is certainly rational to fix the beginning of this process with humans such as Pythagoras and Euclid. These humans represent important benchmarks in the journey from non-scientific beings to scientific beings, and while we may not know all the particulars of how we got to that point, we do have a fair degree of knowledge concerning how our understanding evolved from these individuals to the state it is in today.

Because humans are the only scientific beings on planet earth, some might assume that humanoid beings are the only type of beings in the entire universe that can practice science. In the absence of direct evidence of intelligent alien lifeforms, we cannot test conclusively whether or not this assumption is true. We can, however, examine the forms of knowledge which are already at hand and make a determination whether they support this type of assumption or call it into question. If we cannot rule out the possibility that Pythagorean or Euclidean thinking could evolve in an insectoid or avian species here on earth, then Nimoy's Dictum says that we cannot rule out that kind of possibility on an unknown alien planet either.

While we may not understand how or why birds or various insects have come to do some of the things that they do, one thing we do know is that we are incapable of relating to them in the same way that we are able to relate to other mammalian species such as apes and monkeys, dolphins, dogs, cats, horses, pigs, and so on. There are quite clearly deep and invisible sets of relationships and variables among the discrete orders of life that we don't understand, even though we can perceive their effects and infer their existence. While the evolutionary record points us to a time and state of knowledge where things were "one", we still don't really know how the many came from the one. In real life we see the differences and discontinuities of various species against this background.

For reasons that we do not fully understand, there is a natural differentiation among species, with some being able to directly communicate with each other, and some unable to do so. Beavers are not commonly domesticated like dogs are, and are generally wild creatures out there, doing their own things that have nothing to do with us (as, we must presume, are aliens). But go on the internet and you will find plenty of examples where beavers and humans have entered into domestic relationships. While these seem to be usually the result of anomalous situations (such as when a baby beaver is orphaned and raised by humans), it does show the possibility that exists even though it is not generally accessed.

But it also seems that we cannot domestically relate with all species of animals in this way, although no one can say why this is the case. Generally speaking, no human would think of having a domestic relationship with an insect; yet, domestic relationships can be formed between humans and avians and, in some rare cases, with reptiles such as the king cobra, which will allow humans to interact with it in a domestic way on rare occasions (for instance, some humans are able to "kiss" cobras without being bit by them). When it comes to the alien question, it is this possibility of direct communication with an alien being on the "higher" and "more conscious" level that we operate on that tends to arouse our excitement and interest. This qualifies as a more advanced form of the more primitive domestic relationships that we can form with some other non-human species. The assumption from SETI is that the bare fact of scientific products means the channel of communication must instantly exist; but a careful analysis cautions us not to be reliant on this assumption, but to question it.

If a scientific alien has evolved along a non-mammalian line, for instance, it would be exceedingly arbitrary to assume much of anything about their interests and motivations, just as it would be to assume these kinds of things about insects and avians here on planet earth. In order to make a reliable determination as to whether an insectoid or avian Pythagoras would be an impossible creature within the terrestrial parameters of evolution, we would have to have a firm grasp on the range of factors and conditions that created the human Pythagoras and then show why creatures from non-simian lineages are incapable of reaching the same point. This means that we would have to have prospective forms of evolutionary knowledge permitting us to reliably describe how Pythagoras and Euclid arose from the lower simian order and why the insectoid or avian forms could not reach the same destination. However, as the foregoing discussion has shown, no scientist can describe, with temporal precision, the moment that one discrete form transformed into another, even within the trajectory from lower simian to Pythagoras since we cannot reason our way through and beyond "The Great Leap Forward". Due to the inherent epistemological limitations in our present state of evolutionary knowledge, all we can see from our present vantage point is that a transformation did in fact occur, and that some sort of continuous and active process which we do not have direct access to right now gave birth to these various forms. But "how" that transformation precisely occurred, is not a form of knowledge given to us by our evolutionary science. As stated by Dawkins,

> It makes no more sense (and no less) to aim our historical narrative towards *Homo sapiens* than towards any other modern species—*Octopus vulgaris*, say or *Panthera leo* or *Sequoia sempervirens* . . . if elephants could write history, they might portray tapirs, elephant shrews, elephant seals and proboscis monkeys as tentative beginners along the main trunk road of evolution, taking the first stumbling steps but each—for some reason—never quite making it: so near yet so

far. Elephant astronomers might wonder whether, on some other world, there exist alien life forms that have crossed the nasal rubicon and taken the final leap to full proboscitude.

In this passage, Dawkins warns us not to let our desire to see the evolutionary record turn out to our advantage fool us into thinking it turned out in a certain way for reasons that have anything to do with us. The evolutionary record is not "self-anthropomorphizing" or teleological in a way that is personally significant to humans. If we believe we are seeing "objective" reasons for why evolution turned out in our favor, we can be assured that those reasons are scientifically arbitrary and telling us more about our personal situation as humans, rather than anything objective and scientific about the universe as a whole. In other words, we have no good reasons to think that we are "at the center" of the terrestrial evolutionary process, or that the earth-life system was inherently hell-bent on creating us. And yet, the conventional viewpoint remains precisely that: human consciousness is at the center of consciousness itself and is identical with the whole of all forms of consciousness throughout the universe, instead of being one local subset or variation of a much larger field of consciousness of which we know very little. Lord Kelvin's conceit, it seems, is alive and kicking these days!

An intelligent elephant, for instance, would see its own evolutionary pedigree as the key feature, and ours of lesser significance. We would say the elephant's view was distorted, and they would say the same to us; neither perspective could be said to be right or wrong from the standpoint of cosmological or geological thinking.

While have certainly gained many positive forms of knowledge from the study of evolution, a step-by-step understanding of precisely how the various forms came to be is not one of them. Evolution teaches us, says Dawkins, that "there *must* be a moment in history when there were two animals in the same species, one of whom became the ancestor of all humans and no wombats, while the other became the ancestor of all wombats and no humans". But as to what precisely happened in this "moment", and how precisely it happened, we do not know. This is because the transformation happened in an active set of moments, causally linked in some manner whose precise dynamics are no longer available to direct observation. And while we may feel inclined to assume that the life process we experience today is exactly the same in all times and places, this is simply an inference and may or may not be true. Our inability to actually observe evolutionary transformations in real time, as they occur, places a significant restriction on the precision of our knowledge.

If we consider the attributes and characteristics of some non-human species, we see vague traces of a paradigm-driven behavior that is similar in certain respects to the paradigm-driven behavior of scientific consciousness itself. We can rationally infer from this that these alternative lines of intelligence could blossom into an alien scientific consciousness within the generous potentials of cosmological time. The examples of certain species of bees and birds show that non-human terrestrial consciousness is capable of generating and transmitting geometrical and spatial information. Animals such as apes and dolphins demonstrate a capacity for knowledge that betrays characteristics of the paradigm-driven behavior which is sometimes associated exclusively with human scientific consciousness. Since we don't understand the particulars of the evolution of our own consciousness in relation to these forms, or cannot explain to ourselves why simians alone can breed a Pythagoras or Euclid, we simply cannot make any scientifically reliable declarations concerning what is or is not universal in our own scientific consciousness, or what features of our scientific consciousness would be universally reflected in the mind of an unknown alien one.

Based solely on terrestrial sources of knowledge alone, we must admit that we have no way of knowing at the present if the human understanding of science and mathematics is actually universal or actually idiopathic; our "feelings" that we are doing something universal in nature when we engage in these practices are not enough. In the absence of some sort of objective understanding of the "place" from which the forms of mathematic and geometric truths are derived, we must be cautious when trying to extrapolate human experiences and understandings to unknown alien minds. More importantly, we cannot say anything about how an alien mind would perceive our works, or how we might perceive theirs. Even though the results of scientific and mathematical experience show that something beyond our own subjective imaginings must be at play, we do not know enough about the "objective" realm on its own terms to say just what is subjective, and what is not. Nor can we say with any degree of precision how a non-scientific creature known as Concestor 13 gave rise to Pythagoras and Euclid, or scientifically rule out the possibility that avians, insectoids (such as bees), platypuses, star-nosed moles, or really any other creature that has at least reached a level of structural complexity and operation similar to that of Concestor 13, could not provide an evolutionary platform for a form of alien scientific consciousness on another planet. Nor can we say with any degree of reasonable certainty why the forces of evolution produced simian mathematicians and geometers, as opposed to insectoid or avian ones, or even ones from other lines of the mammalian lineage.[45]

And while we cannot say that the paradigm-related behavior we see in other creatures would be "enough" to one day create a Pythagoras or Euclid along an entirely different evolutionary trajectory, nor can we point to any concrete scientific evidence or reasoning showing why simians, and simians alone, are the only type of species from which a Pythagorean or Euclidean being could emerge within the vast timescales of geological or cosmological time.[46] Moreover, if one of these "alien terrestrial forms" did obtain a scientific consciousness on another planet, we have no idea what kind of scientific world that might be, what kind of scientific products such activities might produce, or just how relatable such a scientific reality would be to us.

The truth is that we have no direct knowledge of the potentialities, capabilities, and characteristics of any minds except those of the humans. And even within what we know about the human mind from our own terrestrial experience we simply do not know why the pathway that manifested actually did so. We do not know if there is some hidden law or logic in the nature of scientific consciousness that made the simian the only form of being that could ripen to the stage of mathematical and geometrical consciousness at this point in the earth's history, or the forms of consciousness that are set to form in the future. Perhaps our present position as earth's

[45] Some conventionalists might say that none of these creatures could become scientists unless they displayed opposable thumbs. Clearly, however, the bare presence of opposable thumbs is insufficient here because not all creatures with opposable thumbs turn into scientists. Similarly, one would also have to demonstrate that no creatures, other than simians, could develop opposable thumbs within the timespan of cosmological or geological time. It is unclear how anyone could make such a showing.

[46] This is a "negative use" of local rationality which "carves spaces" for possibilities of other forms of knowledge to emerge later on.

"signature species" is just some type of strange accident, or temporary aberration, in the longer geological timeframes of this planet. Perhaps there are other planets in the universe where the simians play the fools, and the avians, insectoids, or reptilians rule the roost. Thus, even if there was a law that mandated simian scientific superiority, we do not know what it is. Our ignorance should counsel great caution if we wish to approach the subject matter of alien minds in a realistic, scientific way.

In the absence of direct evidence from alien minds themselves, we are wholly ignorant on the matter because we lack knowledge of sufficient complexity to make the determination as to what precisely such beings are thinking unless we hear it directly from them. Until we are in a position where we can describe with "step-by-step" precision the real process by which non-scientific creatures on earth became scientific ones, we must restrain our imagination from supposing we have enough real knowledge to say anything reliable about alien minds. We should not delude ourselves into thinking that we have any real knowledge of alien minds, when all we are really doing is projecting the things that we would hope to see in them (i.e., that aliens are out there "patiently waiting" to communicate with us or rescue us from all of the problems and dilemmas we have created for ourselves here on earth).

In a similar vein, we should not assume that our understanding of mathematics and science, or even technology, would be universally recognized by other intelligent alien species until some form of actual confirmation has been established. For all we know, there could be scientific alien minds that have arisen from an insectoid line, that perceive space primarily in a hyperbolic manner; or, there could have arisen a scientific alien from a being similar to a duckbill platypus, with a natural inclination towards hyperbolic space and an ability to directly interact with the electromagnetic field through the expression of intention. We simply have no idea just how "alien" such alien minds could really be or how they would relate to the conventional forms of mathematics, geometry, and science as we understand them today. We have no real idea, for instance, if "our" understanding and expression of things like prime numbers or simple arithmetic sums is actually expressing something idiopathic, universal, or something in between—regardless of how universal such expressions may seem or feel to us. Thus, at a minimum, we should not assume that all alien minds would be "just like us", or what their interests, motivations, or proclivities might actually be like.

There can be no doubt that our collective mathematical experience has created an inescapable sense of necessity that things like hyperbolic space are issuing from an independent form of agency even though the discovery of such a thing seems to happen in the subjective inner world of the human mind. While it certainly seems to be the case that mathematics and geometry come from a place of some sort that is wholly independent of whatever we may think about it, since all of our expressions and understanding of that "universal place" are made in real, discrete contexts which are conditioned by our particular evolutionary history and pedigree, we simply lack the perspective and ability to say of much anything about the universal nature of mathematics and science as we understand them vis-à-vis an unknown alien mind. The only way that we could have confidence in the universality of our mathematical and scientific experiences is if we assume that our form of consciousness is "at the center" of all forms of consciousness, no matter how alien they may be. But the lessons from the Copernican Revolution counsel against thinking in these ways and to remain relentlessly focused on contextualizing and specifying the experience of human consciousness to the greatest degree of precision that is possible at the present time.

Thus, while it is certainly *possible* that things such as the Pythagorean theorem, or Euclid's postulates, would be immediately recognizable to an alien mind in the same way that they are for us, we have no way of knowing that this is actually the case until we have some direct feedback from the alien minds themselves. From a standpoint of strict science, therefore, it is arbitrary to conclude, as the SETI paradigm does, that there exists a universal continuity between human minds and all other alien minds in the universe. The arbitrary nature of this assumption means that we should not rely upon it as a solid article of scientific faith to guide future research into these topics, or to exclude fields and avenues of research from consideration. Indeed, it appears self-evident that SETI's assumptions of human to alien relatability are based mainly on an immature emotional desire to connect with a supernatural being rather than a rigorous consideration of real sources of knowledge.

3.5 Terrestrial rednecks

Another related assumption of the SETI paradigm that may not merit much scientific faith is the idea of the "lonely universe"—that is, the idea that space is essentially cold and barren, with remote pockets of life scattered at great distances from each other here and there that never really get to interact with one another. While such an assumption would seem the correct view to a 1950's scientist, we can no longer give this view such great weight in the 21st century. In *Searching*, Morrison and Cocconi noted that they had no reliable way to estimate the number of intelligent alien civilizations that exist in the universe. The same is, of course, true today. However, our picture of what the universe actually looks like has dramatically evolved due to the images from the Hubble and James Webb Space Telescopes.

In 1959, a scientist looking out into the universe could rationally conclude that the universe was probably populated with intelligent life on other planets, but was yet lonely, vast, and spartan. But that is not the type of picture that the universe suggests to us now. Our universe is one that appears to be teaming with life in all corners. Some scientists estimate that there may be as many as 2.5 billion habitable planets in the Milky Way Galaxy alone. This is a staggering number and becomes even more staggering if we were to multiply that number by the number of galaxies that are currently observable, which could be as high as 200 billion. So even though the basic fact of a universe filled with life remains the same from 1959 until today, the content of the fact is far different, and should incline us to naturally take a much different view of the issue today than it would have in the time of Cocconi and Morrison.

Second, there is an underlying assumption in the SETI paradigm that none of these advanced alien civilizations could have reached earth by now. But is it sound to make this assumption? When we consider both the age of the universe, and the modern view that finds our galaxy alone possibly containing 2.5 billion habitable planets, it seems reasonable to conclude that at least some of these civilizations could have reached a level of scientific maturity comparable to us today some time ago. If such a civilization was 500, 1000, 10,000, 100,000, or a 1,000,000 years beyond the point that we are today, can we really say that they could not have made the journey to earth by now? It seems difficult to find such a conclusion determinative in any respect. Even though we are presently incapable of faster than light travel, the work of the mathematician Alcubierre theorizes that faster than light travel is implied in Einstein's field equations. Moreover, many scientists have noted that beings could travel vast distances even within the framework of relativity as presently understood. So we certainly cannot rule out the possibility, on *a priori* grounds, that all intelligent alien civilizations in the universe are incapable of reaching earth today. If the wheel, lever, and wedge, were the most powerful technologies a mere 15,000 years ago, we can perhaps scarcely imagine what our own technological capabilities might look like in another 10,000. Our own experience shows us that technology can take great leaps in what to the eyes of the universe is a very short span of time.

If there are beings in the universe who can travel at galactic ranges within a frame of time relative to their span of life which would allow them to visit a number of different solar systems before their death, then the universe would not be a cold, lonely space to them. We could say that the universe would only appear cold, lonely and barren to terrestrially bound beings, such as ourselves, who lack the ability to easily travel over galactic ranges and for whom the possibility of such travel is nothing more than a science fiction fantasy. But to beings that could actually travel over galactic ranges as easily, as say, I could fly from Seattle to New York, the universe could be a much different place. Thus, we should be very wary of limiting the possibilities of our scientific study of alien anomalies under the hypothesis that all alien civilizations in the universe are only capable of communicating with each other through radio technologies across cold, lonely, spartan distances.

3.6 *The safe distance paradigm's* a priori *rejection of a deep and thorough aerospace search for signs of alien presence appears arbitrary and irrational*

The arbitrary nature of these assumptions in the SETI paradigm means simply that it is unwarranted to limit our study of alien anomalies to deep space radio monitoring. This is not to say that such a search is useless or should be abandoned, as each person has the right to pursue avenues of research that make sense to them. But this is to say that it is short-sighted to only search for aliens in deep space and to assume there is nothing to learn from an aerospace study of alien anomalies. In the early 21st century, we have more than enough direct and anecdotal evidence to justify a substantial and systematic survey of the near-earth environment for alien anomalies. And we should not assume that if aliens reached earth, they would automatically make their presence known. In fact, there may be characteristics and features of these alien minds that would make that possibility either something they have no interest in, or something they would prefer to avoid. Either way, we have no knowledge on the subject matter and should simply leave all options on the table in the spirit of open inquiry on a matter of such great interest and importance.

3.7 *Sagan's dictum regarding extraordinary proof and extraordinary claims is scientifically unsound*

The next element of the SETI paradigm's apparatus of novelty suppression is known as Sagan's dictum which says, "extraordinary claims require extraordinary proof". In this statement, the phrase "extraordinary claim" refers to any claims that aliens are visiting earth. On one level, it certainly would be "extraordinary" for anyone to claim that aliens are visiting earth today and have done so in the past. However, the reason why such a claim would be extraordinary ***is not because it is actually scientifically extraordinary*** but simply because it would be extraordinary from the standpoint of aggregate, normalized human experience.

As noted previously, alien anomalies have been considered "visionary rumors" or outliers from the start of human history. At present, intelligent aliens, or alien civilizations, are nothing more than inferred objects from the standpoint of science, in a manner analogous to the ether of the 19th century and dark matter today. We know nothing about such inferred objects unless and until a proper inquiry has been conducted for a sufficient period of time. Thus, common experience would certainly classify a scientifically-verified alien presence on earth as extraordinary—especially, when aliens are only considered to be "real" in the context of science fiction. But from an objective scientific standpoint, there are no known scientific facts which would make the presence of an alien presence on a planet such as earth at this point in cosmological time any more or less extraordinary because we have no scientific knowledge of the actual distribution of intelligent alien civilizations throughout the universe, and no scientific knowledge about the technologies which some of these alien civilizations could possess. From a standpoint of actual scientific knowledge, it could be the case that the universe is teeming with intelligent alien civilizations with the ability to travel galactic ranges at will. It could also be the case that a species such as humans has been seen many times before, and could possibly be somewhat common, holding no great interest or importance to such alien civilizations than the innumerable hives of bees on earth would be for us. It could also be the case that aliens are visiting earth and regard our presence more akin to the nuisance presented by a colony of ants to someone wishing to enjoy a picnic by the lake. This could be how things "really are" in the universe, even though our human experience to date inclines us to believe something different. But our human knowledge lacks the depth and breadth necessary to make reliable declarations regarding the cosmos as a whole. Indeed, as demonstrated above, even from the mere terrestrial perspective, our knowledge is severely limited when we attempt to rely upon it to derive real alien knowledge.

If we are going to treat alien anomalies as scientific objects, we should not confuse the assumptions generated from ordinary human experience with facts that are scientifically derived. Let us not be like those in the days of Copernicus who rejected the truth of his assertions simply because they did not mesh with the facts of ordinary life as they were known at the time. The case of Copernicus shows that it is sometimes necessary to look beyond the facts of the moment to see a deeper set of relations that are more accurate and precise. In the case of aliens, we have no more scientific facts confirming or denying their presence in earth's aerospace, than the scientists of the 19th century had with respect to the inferred object known as the ether, or that today's scientists have respect to the inferred object known as dark matter. However, aliens today are akin to dark matter—and unlike ether—because we have very rational reasons for wanting to search for them (whereas in the case of ether, the search no longer makes sense after the work of Einstein).

Accordingly, we cannot say from a standpoint of scientific thought itself that there is anything at all which is "extraordinary" about aliens visiting planet earth. While it might be shocking to us from a standpoint of personal history and experience—just as the idea that the earth moved in space was shocking to the regular people of the 16th century—the knowledge we have derived from the common fund of general human experience is insufficient to provide any scientific knowledge whatsoever, and it remains an open field of inquiry to be explored. This last line of thought leads to the second part of Sagan's dictum—the requirement for "extraordinary proof" of alien presence on planet earth. This aspect of Sagan's dictum directly pertains to problems of scientific conversion.

4.0 SCIENTIFIC CONVERSION PROBLEMS

4.1 *Because the study of alien anomalies is in a "pre-scientific" phase, the demand for "extraordinary proof" is intrinsically irrational*

By and large, normal scientific activity does not actively seek anomalies; nevertheless, the very structure of scientific operations tends to induce their appearance. This paradox is explained by the basic nature of normal scientific activity itself. Kuhn likens normal scientific activity to puzzle solving behavior. In order to solve a puzzle, one has to know what pieces one is using, and accept that they are puzzle pieces. If the puzzle pieces don't fit together to make a solution, then the puzzle cannot be solved. In the case of a human-designed puzzle, the user of the puzzle presumes that the solution has been embedded in its design in advance by its creator and that the hidden solution will be revealed if she finds it. The puzzle-solver relies on the advance assumption that a solution exists, even though it may not be apparent for some time while the puzzle-solving activity is underway. If, by contrast, someone purchased a puzzle whose pieces had been deliberately manufactured in a way that made sure they did not fit, then the puzzle could never be solved.

In one sense, the paradigms that guide normal science are analogous to the picture on the puzzle box showing what the completed puzzle is supposed to be like when all the pieces are put together. The difference, of course, is that the scientist does not really know for sure if the picture she has in her mind as she proceeds will ultimately end up being the real picture of things. Yet, she must assume this to be the case to some extent if she wants to have a chance at finding any solution at all. And not only must she adopt a particular viewpoint in advance regarding the form of the solution, she must also commit herself fully to a wide range of practices, disciplines, and other investments, in order to make any headway towards finding the real solution.

> In the development of any science, the first received paradigm is usually felt to account quite successfully for most of the observations and experiments easily accessible to that science's practitioners. Further development, therefore, ordinarily calls for the construction of elaborate equipment,

the development of an esoteric vocabulary and skills, and a refinement of concepts that increasingly lessens their resemblance to their usual common-sense prototypes. That professionalization leads, on the one hand, to an immense restriction of the scientist's vision and to a considerable resistance to the paradigm change. Science has become increasingly rigid. On the other hand, within those areas to which the paradigm directs the attention of the group, normal science leads to a detail of information and a precision of the observation-theory match that could be achieved in no other way. Furthermore, that detail and precision-of-match have a value that transcends their not always very high intrinsic interest. Without the special apparatus that is constructed mainly for anticipated functions, the results that lead ultimately to novelty could not occur. And even when the apparatus exists, novelty ordinarily emerges only for the man who, knowing *with precision* what he should expect, is able to recognize that something has gone wrong. Anomaly only appears against the background provided by the paradigm.

When the scientist narrows her focus, commits to a path of inquiry, and deploys a wide range of investigative modalities, new information and detail emerge. In this process, some of the information may be assimilable into the paradigm that guided the research, some may not; or, new phenomena may emerge which expands the range of paradigm, such as in the case of X-rays, which were not anticipated by Roentgen's research (and rejected at first by Lord Kelvin) but which expanded the range of the paradigm that guided the research instead of shattering it. Paradoxically, the very steadfast commitment to a particular solution also gives rise to new problems, and the first signal that something is not working out is oftentimes presented in some form of anomaly.[47] As the paradigm grinds on and scientific study

---

[47] Anomalies always arise against the background of normal phenomena; normal phenomena, in turn, are always the products of active paradigms. The "form" of the anomaly could be spontaneous or normal. In general, Kuhn identifies three specific kinds of scientific phenomena. The first kind consists of those phenomena "already well explained by existing paradigms . . .". The second consists of phenomena "whose nature is indicated by existing paradigms but whose details can be understood only through further theory articulation. . . . Only when attempts at articulation fail do scientists encounter the third type of phenomena, the recognized anomalies whose characteristic feature is their stubborn refusal to be

deepens, it will either keep successfully solving puzzles, or start creating gaps as its activity discloses the limitations or inaccuracies of the puzzle picture guiding the scientist. Since human-designed paradigms are usually less than totally complete in all possible domains of inquiry, no one should be terribly surprised when gaps start appearing and the paradigm starts to wobble.

If alien anomalies are to be considered as objects of scientific inquiry, then they should be properly placed from the start in an appropriate conceptual category so that they will be viewed and treated in a rational manner. Accordingly, they cannot be considered objects of normal science and subject to the demands of "extraordinary proof" right now because there do not exist any paradigms to explain them; instead, they should be placed in the "pre-scientific" phase of discovery. According to Kuhn, nearly every mature science has passed through a pre-scientific phase before it is properly formed and shifts from into a state of normal scientific activity. Kuhn describes this process as follows,

> . . . the early developmental stages of most sciences have been characterized by continual competition between a number of distinct views of nature, each particularly derived from, and all roughly compatible with, the dictates of scientific observation and method. What differentiated these various schools was not one or another failure of method—they were all "scientific"—but what we shall come to call their incommensurate ways of seeing the world and of practicing science in it. Observation and experience can and must drastically restrict the range of admissible scientific belief, else there would be no science. But they cannot alone determine

---

assimilated to existing paradigms. This type alone gives rise to new theories. Paradigms provide all phenomena except anomalies with a theory-determined place in the scientist's field of vision." Thus, one could say that "normal anomalies" arise in the expected course of scope of normal scientific activity when the scientist understands in advance that they are studying something new. Oxygen and electricity can be placed in this category. Spontaneous anomalies include phenomena such as X-rays or alien anomalies, since they arise out of nowhere against the background of normal paradigms. The Copernican revisions could be seen in this light as a normal anomaly, since Copernicus understood that the "surprising fit" between heliocentric calculations and the observed motions of planetary bodies constituted an anomalous fact against the background of phenomena as normally understood in his time.

> a particular body of such belief. An apparently arbitrary element, compounded of personal and historical accident, is always a formative ingredient of the beliefs espoused by a given scientific community at a given time.
>
> That element of arbitrariness does not alone, however, indicate that any scientific group could practice its trade without some set of received beliefs. Nor does it make less consequential the particular constellation to which the group, at a given time, is in fact committed. Effective research scarcely begins before a scientific community thinks it has acquired firm answers to questions like the following: What are the fundamental entities of which the universe is composed? How do these interact with each other and the senses? What questions may legitimately be asked about such entities and what techniques employed in seeking solutions? At least in the mature sciences, answers (or full substitutes for answers) to questions like these are firmly embedded in the educational initiation that prepares and licenses the student for professional practice.

Dr. Griffin describes this phase of science in another way in relation to the study of animal consciousness in the preface to *Animals Minds* as follows,

> In a sharp break with the traditional conviction that the mental experiences of animals cannot be studied scientifically, some of us have begun to try. This is now timely because ethologists and comparative psychologists have learned so much about the versatility of animal behavior that we can now investigate animal cognition on the basis of extensive evidence not available in the nineteenth century to Darwin, Romanes, or Lloyd Morgan. The difficulties of obtaining convincing evidence remain formidable, partly because the subject has been neglected by the scientists best equipped to investigative it, as discussed by Beer (1992). It is therefore necessary to begin with a variety of intuitions and speculative suggestions that must be tried out in any way that appears promising, in order to learn which, if any, may develop into effective scientific procedures, as discussed by Silverman (1983).

> By necessity, this is "pre-science" which is where scientific advances often begin. Scientists who accept only well-proven methods that can be confidently expected to yield unambiguous data will necessarily be disappointed, because this is unknown territory waiting to be explored, not well-mapped terrain where we know the main features and seek only to pin down specific details. It is not yet possible to formulate cookbook specifications stipulating just how to prove whether a particular animal is conscious. But I hope that those ready to face the challenges of exploring the significant unknown will be stimulated by the prospect of learning what life is life to our evolutionary kin.[48]

The pre-scientific phase lacks coherent paradigms to guide and focus scientific activity. In this phase, the "data" is largely anecdotal, unorganized, circumstantial, unclear, random. But this state of affairs is not necessarily due to any inherent features of the phenomena or anomalies themselves, but is a reflection of the relationship between the phenomena, the active paradigms, and the meaning derived from them by the observing bystander. If scientific activity is continued and later leads to real knowledge, the initial phase will later appear unscientific once the paradigm is developed and scientists "look back". Kuhn contends that this is a necessary phase in scientific development which most (if not all) of the mature modern sciences pass through at one point or another.

> In the absence of a paradigm or some candidate for a paradigm, all of the facts that could possibly pertain to the development of a given science are likely to seem equally relevant. As a result, early fact-gathering is far more nearly random activity than the one that subsequent scientific development makes familiar . . . The resulting pool of facts contains those accessible to casual observation and experiment together with some of the more esoteric data . . . for example, Pliny's encyclopedic writings or the Baconian natural histories of the seventeenth century . . . [are] . . . a morass. One somehow hesitates to call the literature that results scientific. The Baconian "histories" of heat, color,

[48] If one substitutes the terms "alien" for "animal" and "galactic" for "evolutionary" in this passage, it transforms into an example of a scientific attitude towards alien anomalies that is consistent with its status as a pre-science.

> wind, mining, and so on, are filled with information, some of it recondite. But they juxtapose facts that will later prove revealing (e.g., heating by mixture) with others (e.g., the warmth of dung heaps) that will for some time remain too complex to be integrated with theory at all. In addition, since any description must be partial, the typical natural history often omits from its immensely circumstantial accounts just those details that later scientists will find sources of important illumination. Almost none of the early "histories" of electricity, for example, mention that chaff, attracted to a rubbed glass rod, bounces off again. That effect seemed mechanical, not electrical. Moreover, since the casual fact-gatherer seldom possesses the time or tools to be critical, the natural histories often juxtapose descriptions like the above with others, say, heating by antiperistasis (or by cooling), that we are now quite unable to confirm. Only very occasionally, as in the cases of ancient statics, dynamics, and geometrical optics, do facts collected with so little guidance from pre-established theory speak with sufficient clarity to permit the emergence of a first paradigm . . . No wonder, then, that in the early stages of the development of any science different [scientists] confronting the same range of phenomena, but not usually all the same phenomena, describe and interpret them in different ways. What is surprising, and perhaps also unique in its degree to the fields we call science, is that such initial divergences should ever largely disappear.

As noted previously, electrical research in the 18th century spent time in this phase before Franklin's work provided structure and direction to the field. Before Franklin, "there were almost as many views about the nature of electricity as there were important electrical experimenters . . . Yet, though all the experiments were electrical and though most of the experimenters read each other's works, their theories had no more than a family resemblance." If we accept the suggestion that scientific study of alien anomalies today is in a pre-science phase, then we will also conclude that the SETI paradigm's requirement of "extraordinary proof" is asking way too much of a scientific inquiry at such a nascent stage of development. The evidence regarding alien anomalies from 1958 to the present primarily consists of "visionary rumors," with a dash of physical evidence, as well as video and photographic evidence, and anecdotal reports. But nothing scientifically concrete beyond that has been disclosed (at least in the public domain). The type of scientific survey called for in Project Sign in 1947 has not been commenced; nor does mainstream science even recognize that it is worth commencing. In the absence of regular and sustained scientific inquiry, it makes little sense to require *any* substantive scientific proof in advance—let alone "extraordinary" proof. A pre-science phase fact-gathering and analysis is to be expected *before* an expectation of any form of "proof" can even be considered. The idea that extraordinary proof of an alien presence would spontaneously materialize in the absence of sustained and disciplined scientific activity is contraindicated by the scientific record itself.

Moreover, Sagan's admissibility requirement is arbitrary because the claim that evidence of alien presence on earth would be "extraordinary" is not a scientific statement, but rather an expression of personal feelings derived from the conditions of ordinary experience for a species that is unable to travel in space and knows nothing concrete about the actual state of intelligent alien species elsewhere in the galaxy. People of Copernicus' time no doubt found the idea that the earth moved in space "extraordinary", but what did their personal feelings about the facts of their time have to do with actual reality? While adherents of the SETI paradigm have every right to set a high bar of admissibility for their own scientific paradigms, this position is not without consequences, because they will inhibit themselves from engaging in scientific activity based on the arbitrary conviction that extraordinary proof of alien presence would magically fall in their laps if aliens were really here just like it does in the movies.

At the same time, even though SETI's approach to alien anomalies may be scientifically naïve, it does accurately reflect the subconscious yearning of many scientists to "know more" about alien anomalies. But if Sagan's dictum is epistemologically suspect, what would constitute a valid substitution? In other words, if science shouldn't be pounding its fist on the table demanding "extraordinary proof" of alien presence right now, then what, precisely, should it be doing?

4.2 *Scientific conversion processes in general*

Acceptance of the premise that a study of alien anomalies is in a pre-science phase, leads naturally into an examination of the process through which a pre-science phase gets converted into a more normal mode. Kuhn contends that the movement from a pre-science phase to normal science involves much more than simply a reinterpretation of the data derived from normal experimental activities. When a scientist is operating under an established paradigm, she knows "what a datum [is], what instruments might be used to retrieve it, and what concepts [are] relevant to its interpretation. Given a paradigm," Kuhn writes, "interpretation of data is central to the enterprise that explores it."

> But that interpretive enterprise . . . can only articulate a paradigm, not correct it. Paradigms are not corrigible by normal science at all. Instead, as we have already seen, normal science ultimately leads only to the recognition of anomalies and to crises. And these are terminated, not by deliberation and interpretation, but by a relatively sudden and unstructured event like a gestalt switch. Scientists then often speak of the "scales falling from the eyes" or of the "lightning flash" that "inundates" a previously obscure puzzle, enabling its components to be seen in a new way that for the first time permits its solution. On other occasions the relevant illumination comes in sleep. No ordinary sense of the term "interpretation" fits these flashes of intuition through which a new paradigm is born. Though such intuitions depend upon the experience, both anomalous and congruent, gained with the old paradigm, they are not logically or piecemeal linked to particular items of that experience as an interpretation would be. Instead, they gather up large portions of that experience and transform them to the rather different bundle of experience that will thereafter be linked piecemeal to the new paradigm but not to the old.

Interpretation of scientific data is a linear, continuous, step-by-step, and discursive process. It is one side of normal scientific process. Equally important, however, are the "moments" of conversion, instances which are discontinuous, unanticipated, unexpected, intuitive, and gestalt like. These moments have always played an indispensable role in scientific discovery and are features—not bugs—of this process. Classic examples of moments of scientific conversion include Archimedes' "Eureka" moment in his bathtub, Mendeleev's vision of the periodic table of elements[49], Watson's

---

[49] *See* **Sections 5.2.2** and **5.2.3** for an in-depth discussion of Mendeleev's vision.

dream of intertwined serpents revealing the structure of DNA, Newton's falling apple, and similar moments of gestalt scattered throughout the scientific record.

The topic of scientific conversion need not be limited solely to these moments of intuitive experience, however. In fact, one can study the process of conversion in a variety of different contexts and at least arrive at a quasi-informed understanding of this somewhat mysterious scientific experience. For instance, Newton's insight regarding the universal nature of gravity was triggered by a moment in which he witnessed an apple fall to the ground, but that bare insight alone was just a trigger, rather than an entire gestalt which revealed the "universal" nature of gravity all at once. As stated by Bohm in *On Creativity*,

> The movement of insight in which Newton suddenly realized that the moon is falling, even though it never reaches the earth, was evidently quite different from the ordinary process of discursive thought, in which one step follows another more or less logically, over a period of time. Rather, it was an extreme example of something that everybody experiences when he is thinking about a problem containing a number of contradictory or confused factors. Suddenly, in a flash of understanding, involving in essence no time at all, a new totality appears in the mind, in which this contradiction and confusion have vanished. The new totality is at first only *implicit* (i.e., unfolding) through some mental image which, as it were, contains the main features of the new perception spread out before our "mental vision". Perception involving this display, which is inseparable from the act of primary perception itself, is what may be called *imaginative insight* (or creative imagination). Such a display plays a necessary part, because with its aid the mind can apprehend the meaning of what has been created in the flash of understanding. From this apprehension, the mind can go on to think and to reason out more and more of the consequences implied by this new insight.

In the case of Newton, the bare insight into the universal action of falling bodies then led to the question as to precisely how fast these bodies were falling. Newton knew, for instance, that the moon fell more slowly to the earth than an object near the earth surface, but he did not initially have any way to quantify the different rates of fall. As noted previously, Bohm suggests that Newton may have relied upon understandings developed in the study of light to aid in his articulation of the gravitational paradigm.

Similarly, Bohm contends that when Archimedes understood that displacements of volumes of water in his bathtub were unrelated to the shapes of the objects, but directly related to differences in the bodies of the volumes of the objects themselves, it provided a mechanism by which he could "distinguish the specific gravity of different bodies, even though their shapes were too complicated to allow their volumes to be calculated directly from their geometrical proportions." These are variations on the style of analysis presented by Anaximander's fragment, although as time goes on and human observational activities have created more and more narrative orders, the new insights can build upon what has come before. In the case of Archimedes, because the content of the insight itself was not overly complex, the flash of insight likely provided a new conceptual tool which could then be put straight to use. For Newton, however, the essential insight of the falling apple was not enough, by itself, to disclose the law of universal gravitation due to its complexity, but it at least was the spark that set in motion a process by which this insight could be articulated. As Bohm notes, while in the field of specialized sciences this articulation process becomes very technical, at the end of the day this is a faculty that nearly all humans seem to possess, this ability to have an intuitive insight into the order of things and to then parse that insight out with details and measurements so that it can work.

4.3 *The pendulum anomaly*

An example which illuminates some of the challenges inherent to the conversion process is presented by Kuhn in his discussion of how Aristotelian and Galilean physicists would each view the motion of a swinging pendulum. Kuhn discusses the case of the pendulum as follows,

> Since remote antiquity most people have seen one or another heavy body swinging back and forth on a string or chain until it finally comes to rest. To the Aristotelians, who believed that a heavy body is moved by its own nature from a higher position to a state of natural rest at a lower one, the swinging body was simply falling with difficulty. Constrained by the chain, it could achieve rest at its low point only after a tortuous

motion and a considerable time. Galileo, on the other hand, looking at the swinging body, saw a pendulum, a body that almost succeeded in repeating the same motion over and over again ad infinitum. And having seen that much, Galileo observed other properties of the pendulum as well and constructed many of the most significant and original parts of his new dynamics around them.

For each of these physicists the motion of the pendulum is well-explained and non-anomalous. But they have two completely different ways of seeing the motion. If we are looking at these two observers from the position of a present time "meta-observer", it appears to us that Aristotelian can't understand the "correct" explanation of the motion of the pendulum as given by the Galilean. From our perspective, this is an anomaly because as far as we are concerned, the Galilean description of the pendulum should be universally accepted by any rational observer. For us, this appears to be a problem of conversion as we cannot understand why the Aristotelian cannot see the motion of the pendulum the "right way".

To us, the Aristotelian may appear delusional because he is denying the truth of the reality which is literally swinging before his eyes. According to Kuhn, however, the differences between the Aristotelian and Galilean paradigms of motion actually result in these two physicists seeing quite different things when they observe the "same" pendulum moving through space. Thus, we cannot really say that either observer is objectively delusional but only relatively so. What is true, however, is that we (the meta-observers) live in a world where the Galilean paradigm is both superior and closer to the paradigms under which we order our perceptions and thoughts about the laws of motion.

Kuhn notes that Galileo had been trained in "impetus theory", which inclined him to analyze the motion in terms of a "medieval paradigm which held that the continuing motion of a heavy body is due to an internal power implanted in it by the projection that initiated its motion." As a result, when Galileo observed the motion, he was inclined to look for motion which was "symmetrical and enduring". Galileo was also inclined, due to his affinity for Neoplatonism, to be drawn to the "motion's circular form. He therefore measured only weight, radius, angular displacement, and time per swing, which were precisely the data that could be interpreted to yield Galileo's laws for the pendulum." But the Aristotelian would never even consider measuring for such characteristics, and thus would never see them in the motion of the pendulum.

> Seeing constrained fall, the Aristotelian would measure (or at least discuss—the Aristotelian seldom measured) the weight of the stone, the vertical height to which it had been raised, and the time required for it to achieve rest. Together with the resistance of the medium, these were the conceptual categories deployed by Aristotelian science when dealing with a falling body.

As far as the reality of the pendulum is concerned, these two observers live in different worlds. Due to the manner in which Galileo viewed the motion of the pendulum from the start, "pendulum-like regularities were very nearly accessible to inspection." But these regularities were simply not accessible to the Aristotelian regardless of how long he inspected the pendulum. So long as the Aristotelian remained an Aristotelian, the Galilean regularities could simply not materialize within his field of scientific vision.

> Contemplating a falling stone, Aristotle saw a change of state rather than a process. For him the relevant measures of a motion were therefore total distance covered and total time elapsed, parameters which yield what we should now call not speed but average speed. Similarly, because the stone was impelled by its nature to reach its final resting point, Aristotle saw the relevant distance parameter at any instance during the motion as the distance *to* the final end point rather than as *from* the origin of motion. Those conceptual parameters underlie and give sense to most of his well-known "laws of motion."

In essence, the Aristotelian and the Galilean have two different visions of the pendulum. As Kuhn puts it, "the immediate content of Galileo's experience with falling stones was not what Aristotle's had been." For Kuhn, this different vision is not simply a case of two different interpretations of the "same data", but rather two different visions of the world itself. We think that the Galilean's vision is superior because the Galilean's vision is closer to our own and presumably is capable of apprehending both visions in her mind, while the Aristotelian is only capable of a vision which cannot include that of the Galilean. The "true" Aristotelian is so committed to the established paradigm, that he may be unable to free himself from what appears to us as a persistent error of his limited vision of the pendulum, and may even remain unable to apprehend the Galilean vision before his ability to gather knowledge is timed out due to his death.[50] Understanding what needs to be said and done in order to assist the Aristotelian in at least becoming capable of seeing the Galilean motion alongside the Aristotelian one is at the crux of the question of conversion in this example.

In a lecture entitled "The Relations between the History and the Philosophy of Science" (1968), Kuhn approaches this same idea from another angle. Kuhn notes that in its day, Aristotelian physics provided a comprehensive worldview and explanatory structure. Some pieces of this structure contained "lawlike" elements that, if displaced, would obliterate the entire narrative.

> A central tenet of Aristotle's physics was, for example, the possibility of a void. Suppose that a modern physicist had told him that an arbitrarily close approximation to a void could now be produced in the laboratory. Probably Aristotle would have responded that a container emptied of air and other gases was not in his sense a void. That response would suggest the impossibility of a void was not, in his physics, a merely empirical matter. Suppose now instead that Aristotle had conceded the physicist's point and announced that a void could, after all, exist in nature. Then he would have required a whole new physics, for his concept of the finite cosmos, of place within it, and of natural motion stand or fall together with his concept of the void. In that sense, too, the lawlike

---

[50] One of the ways in which paradigm shifts occur is when communities of scientific dinosaurs fail to adapt to the superior paradigm and die off. "When, in the development of a natural science, an individual or group first produces a synthesis able to attract most of the next generation's practitioners, the older schools gradually disappear. In part their disappearance is caused by their members' conversion to the new paradigm. But there are always some men who cling to one or another of the older views, and they are simply read out of the profession, which thereafter ignores their work. The new paradigm applies a new and more rigid definition of the field. Those unwilling or unable to accommodate their work to it must proceed in isolation or attach themselves to some other group." However, science as a whole would benefit more if the Aristotelian converted to the more expansive view while still alive and became an active member of the group of living scientists who continue to deepen their knowledge together.

> statement "there are no voids in nature" did not function with Aristotelian physics quite as a law. It could not, that is, be eliminated and replaced by an improved version, leaving the rest of the structure standing.

In other words, revising the concept of the "void" in the Aristotelian scheme would not be analogous to revising a fact or description *about* the void, but would result in the collapse of the entire Aristotelian scheme itself. It would be similar to what happens when one tries to remove the bottom piece of a Jenga tower late in the game, when one small move near the base of the structure would cause the whole thing to collapse. In other words, there is no way that the Aristotelian can reinterpret the void in a modern way without giving up his entire worldview at once. This can be done, of course, but the exercise is far more involved than a simple conceptual revision. As everyone who has played the game of Jenga knows, there is a huge difference between moving a piece higher up on the structure, where the forces of gravity are not as consequential, and one near the bottom where one slight error can cause the entire structure to fall (and this is, in fact, always the most exciting part of Jenga, when players are forced into a position late in the game where they must remove the most precarious pieces of the tower near the bottom).

4.4 *Clinical perception and paradigm shifts*

A lesser-known example of a change in scientific vision is detailed by Michel Foucault in *Birth of the Clinic: An Archaeology of Medical Perception* (1963). In the opening pages of this work, Foucault describes one of the startling shifts in medical perception which occurred between the middle of the 18th Century and 19th Century as follows,

> Towards the middle of the eighteenth century, Pomme treated and cured a hysteric by making her take "baths, ten or twelve hours a day, for ten whole months." At the end of this treatment for the desiccation of the nervous system and the heat that sustained it, Pomme saw "membranous tissues like pieces of damp parchment . . . peel away with some slight discomfort, and these were passed daily with the urine; the right ureter also peeled away and came out whole in the same way." The same thing occurred with the intestines, which at another stage, "peeled off their internal tunics, which we saw

emerge from the rectum. The oesophagus, the arterial trachea, and the tongue also peeled away in due course; and the patient had rejected different pieces either by vomiting or by expectoration."

Less than a hundred years later, this is how a doctor observed an anatomical lesion of the brain and its enveloping membranes, the so-called "false membranes" frequently found on patients suffering from "chronic meningitis:"

> Their outer surface, which is next to the arachnoidian layer of the dura mater, adheres to this layer, sometimes very lightly, when they can be separated easily, sometimes very firmly and tightly, in which case it can be very difficult to detach them. Their internal surface is only contiguous with the arachnoid, and is in no way joined to it . . . The false membrane are often transparent, especially when they are very thin; but usually they are white, grey, or red in colour, and occasionally, yellow, brown, or black. This matter often displays different shades in different parts of the same membrane. The thickness of these accidental productions varies greatly; sometimes they are so tenuous that they might be compared to a spider's web . . . The organization of the false membranes also displays a great many differences: the thin ones are buffy, like the albuminous skins of eggs, and have no distinctive structure of their own. Others, on one of their sides, often display traces of blood vessels crossing over one another in different directions and injected. They can often be reduced to layers placed one upon the another, between which discoloured blood clots are frequently interposed.

Between Pomme, who carried the old myths of nervous pathology to their ultimate form, and Bayle, who described the encephalic lesions of general paralysis for an era from which we have not yet emerged, the difference is both tiny and total. For us, it is total, because each of Bayle's words, with its qualitative precision, directs our gaze into a world of constant visibility, while Pomme, lacking any perceptual base, speaks to us in the language of fantasy.

The descriptions of medical knowledge from the 18th century almost sound like hallucinations, and it seems hard to believe that these people were actually seeing what they reported; and yet it appears to be the case. The transformation of clinical vision into a form which is relatable to us today was the result of a complex reorganization of medical knowledge which occurred at a variety of different levels. A key component of the shift was the replacement of a paradigm of disease that was abstract from the locus of the manifestation of the disease in the human body and a restructuring of the acts of observation, recordation, and communication concerning the manifestation of disease as it spread and formed in the human body (which is how we logically view disease today). Doctors began paying much more attention to the somatic localization of disease and also began ordering clinical knowledge in a more linear, serialized, and time-dependent fashion. "The clinic," Foucault writes in one place, "owes its real importance to the fact that it is a reorganization in depth, not only of medical discourse, but of the very possibility of a discourse about disease." In essence, the emergence of what we today would consider a "correct" or "clear" perception of clinical medical knowledge was the result of a change in the "non-verbal conditions on the basis of which [clinical medicine] can speak: the common structure that carves up and articulates what is seen and what is said."

Disease is the object of clinical medicine and is obviously a phenomenon which has always been a part of human reality from the start. Disease itself, however, is never directly seen on its own terms, instead we see only its manifestations in the human body. One could say that disease is a persistent human anomaly because it is visibly present, yet difficult and perplexing to understand. Today we can localize and treat many diseases with "cures" that are based on concrete, clinical knowledge. "Disease" as an abstract entity, however, is never directly seen and no rational clinician today would even think about disease in this way. Even today, with our superior medical knowledge, diseases that we do not understand continue to manifest. While the reality of disease presents itself very visibly, and sometimes even violently or forcibly stopping all normal human activities in its tracks, it does not come with a ready-made paradigm of knowledge and understanding. Yet, all humans who experience and see disease do intuitively understand at once that something is happening to a human body which is impairing its ability to function and which, in a sense, is drawing it in the direction of death, instead of in the direction of life. When humans see disease, they feel a natural impulse to want to change its course and bring the functioning of the human body back to a state of balance and life so that we can live and function for a longer period of time. When we experience disease, we know that "something is wrong" without necessarily knowing "what" is wrong. We think that there must be some form of knowledge which can help us treat the disease, and we do possess certain forms of such knowledge today—this implies to us that there is some type of undisclosed paradigm driving the expansion of disease. We can understand how early human beings, with no understanding of human anatomy or physiology, germs, viruses, genetics, bacteria, and the like, must have found disease extremely perplexing and mysterious, and caused exclusively by supernatural beings. Many illnesses and conditions that we can understand and treat today were simply unexplainable and unknowable to them.

In an effort to understand disease, people developed paradigms of medical knowledge. In the 17th and 18th centuries, the dominant paradigm was referred to as "nosology". Clinicians practicing under the paradigm of nosology would look at the body as displaying "signs" to the clinicians of which disease was at work behind the scenes, so to speak. Under the paradigm of nosology, the primary object of clinical science was to understand the species of disease on an abstract level and design the treatment based on the species of the disease that was operating; the "localization" of the disease in the organism, was treated as a "secondary problem." This meant that clinicians of that era were not concerned about understanding how particular disease might be impacting the functioning of, say particular organs or tissues, because how the disease was displaying itself was not the primary fact. The primary fact was understanding, on an abstract level, what disease one was dealing with *before* any treatment was indicated. "Never treat a disease," one celebrated 18th century clinician wrote, "without first being sure of its species."

In *Birth*, Foucault sites to source materials from the architects of the paradigms of nosology to help us understand how they saw the world. For example, one of the architects explained the theory of disease in terms which almost seem supernatural and scientific at the same time.

> The supreme Being is not subjected to less certain laws in producing diseases or in maturing morbific humours, than in growing plants or animals. . . . He who observes attentively the order, the time, the hour, at which the attack of quart fever begins, the phenomena of shivering, of heat, in a word all the symptoms proper to it, will have as many reasons to believe that this disease is a species as he has to believe that a plant constitutes a species because it grows, flowers, and dies always in the same way.

We see again the Anaximandian approach to analogizing orders, but the analogy between plants and disease appears to us to be a "bad fit". In this approach, the patient is an afterthought, rather than the primary focus of concern, which is what we think is important. The clinician in this model is reading through the signs presented by the patient the disease at an abstract level and tailoring the treatment directly towards that abstraction. This approach does not make sense to us. We believe the disease that manifests in the body is the most important thing to study. Foucault describes the strange nosological approach as follows:

> The first disturbance is introduced with and by disease itself. To the pure nosological essence, which fixes and exhausts its place in the order of the species without residue, the patient adds, in the form of so many disturbances, his predispositions, his age, his way of life, and a whole series of events that, in relation to the essential nucleus, appear as accidents. In order to know the truth of the pathological fact, the doctor must abstract the patient. "He who describes a disease must take care to distinguish the symptoms that necessarily accompany it, and which are proper to it, from those that are only accidental and fortuitous, such as those that depend on the temperament and age of the patient."[51] Paradoxically, in relation to that which he is suffering from, the patient is only an external fact; the medical reading must take him into account only to place him in parenthesis. Of course, the doctor must know "the internal structure of our bodies"; but only in order to subtract it, and to free from the doctor's gaze "the nature and combination of symptoms, crises, and other circumstances that accompany diseases".

---

[51] This is a citation from an authoritative early 18th century medical text.

Another example of this alien way of thinking about disease is stated by Foucault as follows, based on citations from authoritative medical texts from early 18th century medicine.

> "The knowledge of diseases is the doctor's compass; the success of the cure depends on an exact knowledge of the disease"; the doctor's gaze is directed initially not towards that concrete body, that visible whole, that positive plentitude that faces him—the patient—but towards intervals in nature, lacunae, distances, in which there appear, like negatives, the "signs that differentiate one disease from another, the true from the false, the legitimate from the bastard, the malign from the benign." It is a grid that catches the real patient and holds back any therapeutic indiscretion. If, for polemical reasons, the remedy is administered too early, it contradicts and blurs the essence of the disease; it prevents the disease from acceding to its true nature, and, by making it irregular, makes it untreatable. In the period of invasion, the doctor must hold his breath, for "the beginnings of disease reveal its class, its genus, and its species"; when the symptoms increase and become more marked, it is enough to "diminish their violence and reduce the pains"; when the disease as settled in, one must "follow step by step the paths followed by nature", strengthening it if it is too weak, diminishing it if it strives too vigorously to destroy what resists it".

This approach may seem insane to us, and precisely the opposite of the way in which clinical medicine operates today. Imagine, for instance, if you went to a doctor's office and he told you he wanted to make you get sicker before he treated you to make sure he knew which disease was troubling you; you would surely consider him mad! When we go to the doctor, the primary focus is to understand what is happening in the human body and to see the manifestation of various symptoms. Once a certain constellation of symptoms is established, the doctor then proceeds to draw on the pool of clinical knowledge to determine which disease may be in operation. But sometimes, disease can overlap and be deceptive. Certain forms of treatment may be applied without result, which then prompts the doctor to reevaluate and assess the situation until the correct fit is found. Under this model, a "new" disease (or new manifestation of a disease) is revealed when a presentation does not fit into existent clinical models of knowledge. When an unknown or new disease is presented, it then becomes an object of clinical study.

Clinical medicine has clearly developed a successful regularity in treating a wide range of conditions and has the ability to absorb and study new and unknown presentations within its normal operations. If someone is sick, a disease is presumed to be present, even if the clinician cannot understand which disease precisely is at play by reference to the existent field of clinical data. But ultimately, the point is to find and deploy the "cure" which will diminish the presence of the disease that is interrupting normal functioning, so that normal functioning can be restored. The data in the modern field of clinical knowledge which serves as the source of this information was not developed as a result of abstract speculation, but rather a long trial and error process involving observations and treatment over a wide range of patients involving a wide range of symptoms. As this data was compiled and ordered, clinical knowledge slowly emerged. When new or unknown presentations arose, the field of knowledge itself changed. This knowledge was not developed by relying exclusively on abstract *a priori* intuitions (as had been the practice under the paradigm of nosology) but concretely over time through the field of clinical observation itself.

In the 18th century, however, there was no base of clinical knowledge which had been developed by a persistent understanding of the effects of disease on tissues, organs, and other structures within the body; instead, all of the clinical knowledge was based on the abstract, "evidence-free" paradigm of nosology. Throughout *Birth of the Clinic*, Foucault quotes from a variety of unknown innovators in the field of clinical medicine who developed the paradigms that we rely on today and which displaced the paradigm of nosology. Many of these innovators were keenly interested in contemplating the art of clinical observation as it relates to the formation of judgments and the applications of theories of disease. Foucault contends that a real shift occurred when clinicians abandoned theories and preconceived notions about what they were going to see and focused their attention on the act of observation itself; in essence, what Foucault describes is a paradigm shift. In the case of disease, this was a significant perceptual shift because to the clinicians of the middle of the 18th Century, disease was mystery whose primary attributes were abstract; the focus for clinicians of that era was understanding the species of disease prior to its manifestation in the body (a view which makes absolutely no sense to us today).

It might be hard to imagine an approach to clinical medicine which did not treat the immediate observation of the disease and its unfolding in the body over time as the primary fact and was more focused on understanding disease in an abstract, theoretical space, which was not intimately connected with the spaces of the human body, yet the paradigm of disease that dominated the field of clinical medicine up until the 19th century, viewed disease in precisely that manner. Apparently, the obsession with an abstract model of disease prevented these clinicians from accurately perceiving what was before them; they were extremely fixated on seeing "signs" instead of taking the time to understand the manifestation of disease in the human body at a level of depth and complexity that was appropriate to the nature of disease itself. This appears similar to the SETI paradigm's obsession with deep space alien megastructures, Dyson spheres, and radio signals; the collective obsession with these fantasy objects, inhibits the SETI proponents from doing the real work of observation, study, and measurement in the aerospace, fooling them into thinking they are being "scientific", when all they are really doing is engaging in science fiction fantasy (just like the nosologists thought they were practicing "real medicine" when they were simply being delusional).

Foucault describes the shift in perception from nosology to the modern clinical mode at one place in his work as follows,

> Observation leaves things as they are; there is nothing hidden to it in what is given. The correlative of observation is never the invisible, but always the immediately visible, once one had removed the obstacles erected to reason by theories and to the senses by the imagination. In the clinician's catalogue, the purity of the gaze is bound up with a certain silence that enables him to listen. The prolix discourses of systems must be interrupted. "All theory is always silent or vanishes at the patient's bedside"; and the suggestions of the imagination—which anticipate what one perceives, find illusory relations, and give voice to what is inaccessible to the senses—must also be reduced: "How rare is the accomplished observer who knows how to await, in the silence of the imagination, in the calm of the mind, and before forming his judgment, the relation of a sense actually being exercised."

Once clinicians were freed from an abstract model of disease, they were able to focus greater attention on what Foucault describes as "a certain qualitative, ever more concrete, more individualized, more modelled refinement; the importance of colour, consistency, texture, a preference for metaphor rather than measurement . . . empirical comparisons and references to the everyday or normal . . .". This reorganization of the perception of disease extended to a different way of looking at the spatialization of disease in the human body as well. Along with a reorientation of the "micro-perceptions" of the clinician's gaze there was a larger reorganization of how perceptions were ordered and presented as clinical knowledge as well.

> ...considered on an overall basis, the clinic appears—in terms of the doctor's experience—as a new outline of the perceptible and the statable: a new distribution of the discrete elements of corporeal space (for example, the isolation of *tissue*—a functional, two-dimensional area—in contrast with the functioning mass of the organ . . .), a reorganization of the elements that make up the pathological phenomenon . . . a definition of the linear series of morbid events . . . a welding of disease onto the organism (the disappearance of the general morbid entities that groups symptoms together in a single logical figure, and their replacement by a local status that situates the being of the disease with its causes and effects in a three-dimensional space) . . . This new structure is indicated—but not of course exhausted—by the minute but decisive change, whereby the question "What is the matter with you?", with which the eighteenth century dialogue between doctor and patient began . . . was replaced by that other question: "Where does it hurt?", in which we recognize the operation of the clinic and the principles of its entire discourse.

Interestingly, Foucault himself ascribes this more concrete, precise, and scientific approach to medicine as the result of a new orientation to disease's relationship to death itself. "Disease breaks away from the metaphysic of evil," he writes, "to which it had been related for centuries; and it finds in the visibility of death the full form in which its content appears in positive terms . . . It is when death became the concrete *a priori* of medical experience that death could detach itself from counter-nature and become *embodied* in the *living bodies* of individuals." The change in clinical medicine did not happen overnight, and it took time to develop the new field of knowledge. Clearly this reorganization in depth was very powerful because it actually reordered the act of perception itself, and changed what people actually saw and reported. Apparently, the fixation on the nosological model of abstract disease, blinded people in a certain sense, drawing them into descriptions of human anatomy and processes which seem like hallucinations when compared to the precise, sequential descriptions that we associate with clinical medicine in our era; this is similar to the ways in which SETI's preoccupation with deep space alien megastructures blinds them to the interesting possibilities of aerospace alien anomalies like the tic tac, and how the fixation on Fermi's "Paradox" or Drake's "Equation" keeps so many stuck in an unproductive doom loop. When the transformation away from this model was complete, and the clinician's fixation on nosology was abandoned, the gaze became a very powerful tool whose value became evident over time.

A similar type of shift must happen in the study of alien anomalies. The fixation on abstract, untestable ideas about "aliens" must yield to a course of study that actually studies what is happening in the aerospace itself. Until the study of alien anomalies breaks free from the metaphysics of cosmic loneliness and the subterranean homesick alien, and simply becomes a descriptive enterprise with the goal of gaining greater visibility into the anomalies themselves, it will remain unscientific, speculative, and unproductive, frozen in the same place it has been since the time of Jung (just as the field of clinical medicine was frozen in stasis until the shift away from nosology).

4.5 *The unique conversion of Helen Keller*

An interesting example of a conversion process occurring in a completely different context is raised by Bohm in his discussion of Helen Keller. As is well known, Ms. Keller was a 19th century American who lost her sight and hearing when she was less than 2 years old due to an unknown illness. For several years, she was able to communicate with those in her immediate family through the use of rudimentary "home signs", but her ability was limited. When she was approximately 7 years old, she was brought under the tutelage of a woman named Anne Sullivan. Ms. Sullivan was able to give Ms. Keller her first understanding of concepts by showing her what "water" was. Bohm describes the conversion process by which Keller learned her first concepts in *On Creativity* as follows,

> The key step here was to teach the child to form a *concept* (which she had never learned, because she had not been able to communicate with other people to any significant extent). This was done by causing her to come into contact with water in a variety of different forms and contexts, each time scratching the word "water" on the palm of her hand. For a long time, the child did not grasp what it was all about. But suddenly she realized that all these different experiences referred to one substance in many aspects, which was symbolized by the word "water" scratched in her palm. This initiated a fantastic revolution in the whole of her mind, the depth and scope of which we find hard to appreciate without having experienced directly what it means to live without conceptual abstractions. As a result, where there had been a child without the ability to communicate or even to think, there was now more or less a normal human being.

In her autobiography, *The Story of My Life*, Keller described this moment of conversion as follows,

> I stood still, my whole attention fixed upon the motions of her fingers. Suddenly I felt a misty consciousness as of something forgotten—a thrill of returning thought; and somehow the mystery of language was revealed to me. I knew then that w-a-t-e-r meant the wonderful cool something that was flowing over my hand. The living word awakened my soul, gave it light, hope, set it free!

Prior to this "moment" of conversion, Keller was only able to note a series of differences among phenomena without understanding that there was a similar structure to each of them; the veil of ignorance wasn't lifted until the moment when the concept of "water" was formed in her mind. "This understanding," Bohm writes, "was not merely the result of what she had known before, nor was it even merely the perception of a new relationship in orders that she had already encountered before. It was, in fact, the first perception of an entirely new order in the mind: the order of the concept."

We cited this example because it shows in a dramatic and simple way how new orders of knowledge can arise in a way that is sudden, magical, and revelatory at times. We believe that the paradigms of knowledge in the field of alien anomalies may appear to us in a way similar to manner in which Ms. Keller suddenly apprehended the notion of the concept. In this moment, a new field of knowledge suddenly appeared; facts which seemed isolated and random suddenly cohered in a structured way which opened up new field of exploration. We believe that those scientists who think that all we need is "more data" in order to apprehend alien anomalies in a clear light are mistaken. We believe that dramatic and profound shifts of a nature similar to that of Helen Keller will likely be the heralds of new domains of alien knowledge, and we look forward to the time when those shifts actually occur and take root, leading to the growth of new domains of knowledge.

4.6 *Scientific conversion and the "discovery" of oxygen*

A different kind of conversion process occurs when scientists within a normal field come to see the facts which they have been studying in a completely different conceptual light. A paradigmatic example of this activity is revealed in the case of the discovery of oxygen in the 18th century. Kuhn notes that by the early 1770's at least three chemists had enriched air in a laboratory vessel without knowing it. The first was a Swiss apothecary named C.W. Scheele. But Scheele did not publish his findings before the "formal" discovery of oxygen was widely announced and, according to Kuhn, "had no effect on the historical pattern" of oxygen's discovery. The second discoverer of oxygen was the British scientist Joseph Priestly, who in 1774 collected oxygen as it was released by the burning of red oxide of mercury. The third was the French chemist Lavoisier who reported that the gas released by burning red oxide of mercury was "air itself entire without alteration [except that] . . . it comes out more pure, more respirable." By 1777, as a result of the dialogue between Lavoisier and Priestly, Lavoisier concluded that his "air" was a distinct species of gas, a conclusion with which Priestly disagreed, as Priestly considered his discovery to be "dephlogisticated air" (a term that has no meaning in active science today) but not a distinct species of gas. Kuhn describes the conundrum raised by the "discovery" of oxygen as follows,

> Was it Priestly or Lavoisier, if either, who first discovered oxygen? In any case, when was oxygen discovered? In that form the question could be asked even if only one claimant had existed . . . Priestly's claim to the discovery of oxygen is based upon his priority in isolating a gas that was later recognized as a distinct species. But Priestly's sample was not pure, and, if holding impure oxygen in one's hands is to discovery it, that had been done by anyone who had ever bottled atmospheric air. Besides, if Priestly was the discoverer, when was the discovery made? In 1774 he thought he had obtained nitrous oxide, a species he already knew; in 1775 he saw the gas as dephlogisticated air, which is still not oxygen or even, for phlogistic chemists, a quite unexpected sort of gas. Lavoisier's claim may be stronger, but it presents the same problems. If we refuse the palm to Priestly, we cannot award it to Lavoisier for the work of 1775, which lead him to identify the gas as the "air itself entire". Presumably, we wait for the work of 1776 and 1777 which led Lavoisier to see not merely the gas but what the gas was. Yet even this award could be questioned, for in 1777 and to the end of his life Lavoisier insisted that oxygen was an atomic "principle of acidity" and that oxygen gas was formed only when that "principle" united with caloric, the matter of heat. Shall we therefore say that oxygen had not yet been discovered in 1777? Some may be tempted to do so. But the principle of acidity was not banished from chemistry until after 1810, and caloric lingered until the 1860's. Oxygen had become a standard chemical substance before either of those dates.

The core problem in retrospectively asking "when was oxygen discovered?" arises because the substance that we today think of as "oxygen" was discovered before a paradigm came into being explaining what that substance really was (or, more precisely, what that term means to us today as the ones who are asking the question). We can accurately say that both Priestly and Lavoisier were pioneers in the field of chemistry and that their work paved the way for the discovery of oxygen. However, what we mean today when we use the word "oxygen" represents a form of knowledge that neither of these individuals could possess at the time they did their work. "Ignoring Scheele," Kuhn writes, "we can safely say that oxygen had not been discovered before 1774, and we would probably also say that it had been discovered by 1777 or shortly thereafter. But within those limits or others like them, any attempt to date the discovery must inevitably be arbitrary because discovering a new sort of phenomenon is necessarily a complex event, one which involves recognizing both *that* something is and *what* it is." The case of oxygen represents a situation in which the presence of the "new" scientific object is established before an adequate set of paradigms has come into play which can explain to the scientific community what that new object "actually is". In these cases, there is a delay or lag time between the moment the new object is recognized as being actual, and the paradigms which will situate that new object within a world of scientific meaning and exploration. We apply this example to the study of alien anomalies as follows: we may establish the "real presence" of alien anomalies well before we have any real idea what they really are or what we are really dealing with and must be content with such a temporary state of ignorance and resist the urge to engage in excessive speculation or ideation, while continuing productive work activities.

4.7 *Scientific conversion and alien anomalies*

Put differently, there is a distinct possibility that a scientific study of alien anomalies might be limited to merely measuring and recording alien anomalies for quite some time before anyone has any idea what is really happening. Cases like the discovery of oxygen demonstrate the importance of holding an open period of scientific exploration without the need for "instant understanding". Furthermore, the development of clinical medicine shows us that if theories pile up too quickly and strongly, and the observing bystander becomes overly attached to them, they may be blinded to the nature of the anomalies themselves; thus, it is important (and especially so here) to resist the impulse to "understand" what we are seeing. In some cases, the understanding will not arise for a period of time, and one simply has to learn to be comfortable with the uncertainty that prevails and resist the urge to jump to firm conclusions because those conclusions may distort the possibilities of actual perception in an arbitrary way; instead, we must initiate a scientific process and wait for the fruits of conversion to ripen.

Throughout his discussions on scientific discovery and conversion, Kuhn urges the reader to be mindful of the fact that misaligned attitudes about the scientific discovery process itself can lead us astray. In the case of oxygen, for instance, one may think that one can find a precise answer to the question of "when was oxygen discovered?", but a closer look at the scientific record itself reveals that such a question cannot actually be answered in the way one would like. "Discovery is not the sort of process about which the question is appropriately asked. The fact that it is asked—the priority of oxygen has been repeatedly contested since the 1780's—is a symptom of something askew in the image of science that gives discovery so fundamental a role."

Paradoxically, an unhealthy focus on "scientific discovery" may actually inhibit a scientist from doing the type of work that paves the way for exotic anomalies to be properly understood and worked with. Kuhn describes this as follows with respect to the paradigm shifts that the discovery of oxygen occasioned.

> Only when all of the relevant conceptual categories are prepared in advance, in which case the phenomena would not be of a new sort, can discovering *that* and discovering *what* occur effortlessly, together, and in an instant.
>
> Grant now that discovery involves an extended, though not necessarily long, process of conceptual assimilation. Can we also say that it involves a change of paradigm? To that question, no general answer can yet be given, but in this case at least, the answer must be yes. What Lavoisier announced in his papers from 1777 on was not so much the discovery of oxygen as the oxygen theory of combustion. That theory was the keystone for reformulation of chemistry so vast that it is usually called the chemical revolution. Indeed, if the discovery of oxygen had not been an intimate part of the emergence of a new paradigm for chemistry, the question of priority from which we began would never have seemed so important. In this case as in others, the value placed upon a new phenomenon and thus upon its discoverer varies with our estimate of the extent to which the new phenomenon violated paradigm-induced anticipations.

The "lag time" which sometimes occurs between the moment an anomaly starts to transform into a phenomenon within a network of intelligible paradigms, and the moment at which the anomaly actually becomes a standardized phenomenon within the realm of normal scientific activity, touches upon the paradoxical interplay between anomalies and the new paradigms which these anomalies can spawn. The conventional viewpoint of this process is that the transformation is a result of a change in "the scientist's interpretation of observations that themselves are fixed once and for all by the nature of the environment and the perceptual apparatus." Kuhn strongly disagrees with the conventional viewpoint on the grounds that there is no such thing as an "objective" interpretation and perception of environmental data for a human being; all perceptions and cognitions are filtered through paradigms which have been built up over a long period of time, since before the days of Euclid, Pythagoras, and their ilk. How and why these first paradigms initially formed remains a mystery; however, the fact of their existence, operation, and ongoing transformation all present stable objects of study and analysis.

Kuhn invites us to reconsider the conventional viewpoint, which he argues is itself actually "an essential part of a philosophical paradigm initiated by Descartes and developed at the same time as Newtonian dynamics. That paradigm," Kuhn says,

> . . . has served both science and philosophy well. Its exploitation, like that of dynamics itself, has been fruitful of a fundamental understanding that perhaps could not have been achieved in another way. But as the example of Newtonian dynamics also indicates, even the most striking past success provides no guarantee that crisis can be indefinitely postponed. Today research in parts of philosophy, psychology, linguistics, and even art history, all converge to suggest that the traditional paradigm is somehow askew. That failure to fit is also made increasingly apparent by the historical study of science to which most of our attention is necessarily directed here.

An alternative to this conventional view of discovery is the notion advocated by Bohm and others that science is not a search for the truth, but rather a process by which our perceptual and cognitive engagement with the world is deepened and expanded. "Science is *mainly* a way," Bohm writes in *The Special Theory of Relativity*, "of extending our perceptual contact with the world, rather than of accumulating knowledge about it . . . scientific research does not lead to absolute truth, but rather (as happens in ordinary perception) an awareness and understanding of an ever-growing segment of the world with which we are in contact." Put differently, Bohm says, science is "a perceptual enterprise, rather than . . . an accumulation of knowledge." Under this view, paradigms are used as crutches to make things "fit" in the world, understanding that one day the crutch may no longer be effective or needed. Instead of looking for "true" paradigms, one looks for paradigms which are as accurate and precise as possible under the conditions in which one is presently operating. A paradigm is used until it has been falsified, at which point it may be discarded or modified for a paradigm that is a better fit with the conditions at hand. The conditions in which a human finds him or herself are always subject to change, and no one ever truly believes that they have much of anything figured out for certain.

Under the conventional model, it is presumed that the scientist is getting direct information or data from reality itself; under the modern view, it is presumed that all data and information is filtered by paradigms and processes of which the scientist is never fully conscious or aware. It is this continuing process of deepening and developing an awareness of the limits, conditions, and restrictions upon understanding and perception, displayed in relationship to an actively changing and mutating environmental milieux, that gives rise to greater orders of scientific knowledge. This process is hyperbolic in the sense that it never ends, and one never reaches the boundary line that tells the "ultimate" truth.

> I am . . . acutely aware of the difficulties created by saying that when Aristotle and Galileo looked at swinging stones, the first saw constrained fall, the second a pendulum. The same difficulties are presented in an even more fundamental form by the opening sentences of this section: though the world does not change with a change of paradigm, the scientist afterward works in a different world. Nevertheless, I am convinced that we must learn to make sense of statements that at least resemble these. What occurs during a scientific revolution is not fully reducible to a reinterpretation of individual and stable data. In the first place, the data are not unequivocally stable. A pendulum is not a falling stone, nor is oxygen dephlogisticated air. Consequently, the data that scientists collect from these diverse objects are, as we shall shortly see, themselves different. More important, the process by which either the individual or the community makes the transition from constrained fall of the pendulum or from dephlogisticated air to oxygen is not one that resembles interpretation. How could it do so in the absence of fixed data

> for the scientist to interpret? Rather than being an interpreter, the scientist who embraces a new paradigm is like the man wearing inverting lenses. Confronting the same constellation of objects as before and knowing that he does so, he nevertheless finds them transformed through and through in many of their details.

The case of alien anomalies is unlike any other scientific conversion problem that human beings have ever faced, due to the nature of the "objects" themselves, and the goals of the scientific inquiry (i.e., to establish a "channel of communication" with an active alien intelligence). Science has never attempted anything of this nature before. Although it is reasonable to take issue with the arbitrary limitations of SETI's approach, no one can disagree with the ultimate goal of SETI's study, which is to establish some form of conscious connection with an active alien intelligence. To date, all of the objects of scientific study have either been natural processes, inert objects, or terrestrial non-human species. Humans have never attempted to study beings who presumptively operate from a higher order of knowledge than we do, and who presumptively can perceive, think, and act in ways that are essentially unknowable to us.

We believe that one of the key reasons why progress in this alien field of inquiry has been stymied for so long is due to the subconscious attitudes that the scientific community brings to bear on this subject. Our hypothesis is that the technological capabilities to measure, record, and understand these anomalies are being vastly underutilized because those who are in a position to develop and deploy these technologies simply don't understand that they are not dealing with the normal scientific processes and procedures that they were trained in and have come to expect. Only after they step back and re-examine their suppositions and beliefs will they see the path forward.

Processes related to scientific conversion do not need to be deeply considered in the processes of normal scientific activity. But when one is presented with an anomaly that has no explanatory paradigm, cannot be reliably measured, and whose mere ontology is subject to constant debate (as in the case of alien paradigms), considerations of this nature are critical. Having an understanding of the means and ways by which revolutionary anomalies were "converted" into more normal science in prior eras will provide necessary guidance into new and unknown regions of knowledge.

4.8 *Clarke's Dictum and the scientific conversion of alien anomalies*

Arthur C. Clarke's dictum states as follows: "Any sufficiently advanced technology is indistinguishable from magic." In the context of Kuhn's analysis of scientific revolutions, the "magic" refers to a process which we believe is real, which we believe is happening, but which we cannot understand; thus, at a certain level, we tend to doubt the truth of what our experience is telling us. However, the phrase "sufficiently advanced technology" is a statement indicating that we know that the "magic" we are witnessing is real and must be a result of an articulable scientific process. Even if we cannot understand "how" the magic is happening, and even though it confuses, perplexes, and bewilders, since it appears to be the result of some exotic technology whose products we can witness, we are possessed of a basic belief that the process is articulable and understandable. We hope that, if an alien intelligence can develop and deploy such a technology, perhaps we might one day too.

Clarke's dictum serves as a possible restatement of the paradoxes of scientific conversion in the context of alien anomalies; for although they may seem unbelievable, if they are based on scientific processes, then they must somehow be real to us in a way that we could potentially understand, but at present do not.[52] A contemplation of Clarke's Dictum in this context is also a productive substitution for the fixation on disclosure.

---

[52] In the context of scientific conversions, Kuhn suggests that the scientists attempting to understand "foreign paradigms" must adopt a different mindset towards the unknown paradigm if it is shown to produce real results. "If the new viewpoint endures for a time and continues to be fruitful," Kuhn writes, "the research results verbalizable in this way are likely to grow in number. For some men such results alone will be decisive. They can say: I don't know how the proponents of the new view succeed, but I must learn; whatever they are doing, it is clearly right." This could be a statement of how a human might approach the "magic technology" of aliens if they were to witness it.

On one level, disclosure refers to the disclosure of information held by various governmental and military agencies about information and evidence they have gathered on alien anomalies. In America, for instance, many people believe that their government has collected a vast amount of data and evidence regarding alien anomalies which it is shielding from the public and which, if revealed, would at least settle the question as to whether or not aliens have been visiting earth. There is, of course, good reason for this belief given the fact that the U.S. Military would be the most likely institution to directly encounter alien anomalies and have the technological means and resources to capture data and information about those encounters. Furthermore, we know that the U.S. Government was taking alien anomalies seriously back in 1947 with Project Sign. Thus, it seems likely that the U.S. Government has at least some valuable information that it could share with the general public. Yet, for reasons that are not entirely clear, it steadfastly refuses to do so, prompting many to claim that "disclosure" of this information and data should occur and would be scientifically valuable.

Another aspect of disclosure concerns the discovery or confirmation of alien presence that may occur through non-military scientific activities. There are currently a small number of scientific groups working on studying alien anomalies, with a goal to convincing mainstream academic scientific institutions that the topic is worthy of further attention. These groups are mainly composed of academics and formerly military personnel. Some of these groups are engaged in building observational apparati, direct studies, and attempts to publish peer-reviewed papers. However, all of these efforts share one thing in common from the perspective of scientific revolutions; namely, that they operate within closed, segmented groups of scientific specialists.

The deep structural parallels between the Copernican Revolution and the scientific study of alien anomalies arise because in both cases the scientific facts constitute ordinary facts of life occurring in the real world and are not restricted to the specialized observational apparati of scientific specialists. This is a special and rare type of case when it comes to the subject of scientific revolutions. Normally, the work of specialized segments of the scientific community are immune to what happens outside of their closed circles of knowledge. As Kuhn writes in an essay entitled, "The History of Science" (1968),

> The practitioners of a mature science are men trained in a sophisticated body of traditional theory and of instrumental, mathematical, and verbal technique. As a result, they constitute a special subculture, one whose members are the exclusive audience for, and judges of, each other's work. The problems on which such specialists work are no longer presented by the external society but by an internal challenge to increase the scope and precision of the fit between existing theory and nature. And the concepts used to resolve these problems are normally close relatives of those supplied by prior training for the specialty. In short, compared with other professional and creative pursuits, the practitioners of a mature science are effectively insulated from the cultural milieu in which they live their extraprofessional lives.

Of course, this general rule did not apply in the case of the Copernican Revolution because the position of the earth was not a fact confined to the laboratory; nor is it the case that it applies to alien anomalies, which are similarly not confined. Drawing on this analogy, we recall that the Copernican Revolution did not occur until Copernicus' hypotheses were democratized and concretized by the introduction of the telescope, which allowed people to "see for themselves" that what Copernicus was saying was probably true. We know that the revolution did not become real until that point in time, because the suppression of the anomaly did not become overt until the work of Galileo. The same type of social conversion process may be required in the case of alien anomalies. Even if the closed segments of scientists in the modern world may view themselves as holding the power to change people's minds with their declarations, the real world may not actually function like that and their power may be more limited the further it ventures from the closed, segmented circles in which they principally operate. Within those circles, a certain form of discourse regarding precision, measurement, theory, and fit occurs among those who have similar levels of knowledge. *See e.g,* Kuhn, "Energy Conservation as an Example of Simultaneous Discovery" and "The History of Science", in *Essential Tension.* When we go outside of the circle, the requirements for conversions regarding facts which were established inside the closed segmented circle may change.

So, for instance, if a group of scientists tells the rest of the world that they are viewing certain results in a bubble chamber, particle collider, or a device such as the Laser Interferometer Gravitational Observatory ("LIGO"), the public has no basis to tell the scientists they are wrong, and their *dictat* on the subject should be sufficient. But not all facts arise within closed laboratory settings. The belief that disclosure from segments of the scientific community or government agencies will be sufficient to change perceptions or create new scientific facts may be misplaced, since alien anomalies exist in the world at large (in other words, a scientist's statement about whether aliens were "real" or not, based on the results of some type of specialized investigation that was not a scientific spectacle that observers could witness for themselves, still wouldn't change the fact that aliens are not landing on the White House lawn and demanding to see the President).

But the concept of disclosure in the context of alien anomalies operates at another level as well. In this field, there is a widespread belief, held by many, that there will be a sort of "alien revelation" of new knowledge in a profound and dramatic act of disclosure. It is believed that once this "disclosure event" occurs, nothing will ever be the same again. People yearn and wait for this "disclosure", hoping it will come one day and constantly predicting that this event is just around the corner.

In **Section 5.2.2**, we will discuss the scientific spectacle of Otto von Guericke's "Madgeburg hemispheres". This serves as an example where scientific knowledge appears like magic to an ignorant audience. In that case, von Guericke demonstrated the existence of the vacuum to an 17th century audience who had absolutely no idea that a vacuum even existed. To the ignorant audience, it seemed like magic; but to von Geuricke, it was just scientific fact. Today, it is a very well-understood phenomenon that wouldn't surprise anyone who had a basic understanding of science.

This shows us that dramatic "revelatory effect" is as much a function of relative paradigm perspective as anything else. At a certain level, the anticipation and expectation of a revelatory effect in the context of alien anomalies makes perfect sense because we are childish fools in relation to an alien species that can travel over a galactic range and work with time, space, and matter, on advanced levels. But we should by no means therefore relegate ourselves to this position forever. We should be like an interested member of von Geuricke's audience who goes up to him and asks him to explain the scientific process behind his demonstration, instead one of the wide-eyed yokels in the audience with mouth agape.

It is suggested in this section that when we engage with alien anomalies from the perspective of Clarke's Dictum, we will be less enamored of the spectacle, and more interested in the process that makes it possible.

In this connection, viewing disclosure as a revelatory event is contrary to the how the process of scientific conversion actually works in these types of cases. In situations where an individual scientist is working on a specific problem, of course, a gestalt moment of scientific revelation may occur. But real scientific conversion is not a "one off" event, but rather a process that generally goes through a very unsatisfying and random "pre-science" phases before the clarifying paradigms begin to emerge. Analogizing to the case of the Copernican Revolution, we can see that the nature of alien anomalies are so stunning and run so counter to the truths of ordinary experience, that a "counter experience" may be necessary to assist individuals in the conversion process, and we may not be able to solely rely upon the "word of the scientist". In the case of the Copernican Revolution, the counter experience was provided when people looked through Galileo's telescope and saw, among other things, the moons of Jupiter. There is a certain magical quality in these moments of conversion that seems to break the strangehold of conventional thought and calls to mind Clarke's dictum. Without a counter experience of this nature, it is unclear how conversion from the conventional view might occur. But the counter experience, or scientific spectacle, is not the end result; it is better thought of as a gateway that then opens our minds and inspires us to start learning about the process that made it possible.

4.9 *The strange case of Oumuamua*

Kuhn's work shows that the goal of scientific inquiry is to turn anomalies into workable scientific facts; however, that does not happen until a coherent paradigm has been articulated, allowing groups of scientific communities to work with those facts in a coherent manner. If this is true, how does modern scientific discourse at least begin to build a paradigm of communication that would allow for the possibility of alien anomalies to one day be understood as "normal" scientific facts? We know from our recent experience that the present structure of scientific discourse itself is simply not able to work with alien anomalies in a manner that would even lead to a regular, normalized discourse about alien anomalies. Evidence of this is shown by the fact that our understanding of alien anomalies has barely advanced at all since the descriptions that Jung provided in 1958, whereas our knowledge of things such as computer science, nuclear physics, and medicine has advanced so much farther.[53] Evidence of this can also be seen in the strange case of Oumuamua. *See* Avi Loeb, *Extraterrestrial: The First Signs of Intelligent Life Beyond Earth* (2021).

Oumuamua constitutes the first interstellar object ever observed by humans. Its advocate for its status as an alien anomaly was Dr. Avi Loeb, a well-recognized authority in the field of astrophysics. Oumuamua entered the orbital plane of our solar system on September 6, 2017, from a trajectory which appears to have originated in the direction of the star Vega, twenty-five light years away. It appeared in the Local Standard of Rest. Oumuamua reached perihelion on September 9, 2017, and then made its exit from our solar system in the direction of the constellation Pegasus on or about October 7, 2017.

Oumuamua received its classification as the first interstellar object (1I/2017) in November of 2017. During Oumuamua's passage through our solar system we collected direct data that leads to inescapable inferences regarding its characteristics. Oumuamua's brightness was measured by different telescopes during a 3-day period in October of 2017. The amount of reflected sunlight from this object varied periodically by a factor of ten as the object rotated every 8 hours. This data implies an extreme shape that is about 5x to 10x longer than wide and was likely a few hundred yards by a few tens of yard in length. This profile is exotic by definition because it indicates "geometry more extreme by at least a few times in aspect ratio—or its width to its height—than the most extreme asteroids or comets that we have ever seen."

---

[53] When we speak of "advancing our understanding" of alien anomalies, we are not presuming that the cause of any of the anomalies are aliens, but simply stating that scientific knowledge should be able at some point to provide a *concrete* answer to the ultimate causes of at least some of the anomalies—instead of simply saying "unidentified"—whatever that cause may really be.

There are other exotic aspects to this object as well. Astrophysicists other than Dr. Loeb have concluded that the likelihood of Oumuamua being disc-shaped was 91 percent. The object failed to show any signs of a hot surface temperature, despite its proximity to the Sun. Oumuamua also possessed an unusual luminosity, "at least ten times more reflective than typical solar system asteroids or comets." Lastly, Oumuamua's trajectory from our solar system could not be accounted for solely based on natural gravitational forces. No evidence of any "rocket effect" or devolatilization was revealed during the Oumuamua's passage, even though this object was sufficiently observed from a number of points to collect evidence of these occurrences. As stated by Dr. Loeb, this object "was a small, oddly-shaped, shiny object that deviated from an orbit shaped by the Sun's gravity without showing any discernible cometary tail (caused by the outgassing of a comet's ice turning to steam by friction and the warmth of the Sun) despite a deep search for it with the Spitzer Space Telescope and other detectors."

Dr. Loeb hypothesized that Oumuamua is a "new class of thin interstellar material either produced naturally . . . or is of artificial origin . . . one possibility is that Oumuamua is a lightsail, floating in interstellar space as debris from advanced technological equipment."[54] However, scientific consensus presently classifies Oumuamua as a natural object. In a recent paper entitled *A Study of an Interstellar Object Explorer (IOE) Mission*, Planetary and Space Science 241 (2024), for example, the authors refer to Oumuamua with certainty as a natural object ejected from a planetary system. However, an examination of this position shows that Oumuamua's status as a natural object is no more plausible than Loeb's alien lightsail hypothesis.

The ongoing debate regarding the status of Oumuamua centers around two sets of its characteristics which are not in dispute. First, Oumuamua's trajectory through our solar system was not solely a function of the inherent gravitational forces of planetary bodies. Oumuamua displayed a non-gravitational acceleration on its way out of our solar system. The only known natural mechanism which

---

[54] Loeb was criticized for suggesting the alien lightsail hypothesis because he was working on a project involving lightsails at the time, thus implying that he had "lightsails on the brain". But this is a silly criticism because scientists routinely draw on analogies from other work they are doing, or other aspects of life, to come up with some of their best ideas.

can cause such a change in momentum is the cometary "rocket effect". This leads to the second undisputed point: no rocket effect was observed by the Spitzer Space Telescope as Oumuamua passed through our range of perception. As a result, the proponents of the naturally occurring object theory for Oumuamua have attempted to posit unknown and unseen natural objects to explain its appearance. Some of the proposals that were quickly dismissed were things like "cosmic dust bunnies". One idea that got a bit of traction was the hypothesis that Oumuamua was a cometary body containing frozen molecular nitrogen. This theory is plausible on its face because the off gassing of molecular nitrogen would not have been visible to the Spitzer Space Telescope as 'Oumuamua passed through our solar system.

However, the odds of Oumuamua being composed of frozen molecular nitrogen, or indeed any naturally occurring object, are absurd. In a February 2017 paper entitled *An Observational Upper Limit on the Interstellar Number Density of Asteroids and Comets*—authored in part by Robert Jedicke, a member of the ISSI Team that first detected Oumuamua and continues to strongly argue it is a naturally occurring object—the authors state the following:

> Given our most reliable calculation of the expected interstellar ISO number density it would be essentially impossible for Pan-STARRS1 to ever detect ISOs barring a statistical fluke.

In other words, the laws of statistics, and our best understanding of mainstream astronomy at the time of the sighting, indicate that a natural interstellar object should have never been observed by the Pan-STARRS1 telescope to begin with. The authors of this paper also declare the detection of the first "non-microscopic ISOs will require new survey systems like the LSST . . . and some luck," and that even with LSST's vastly superior surveying power it would not detect "any ISOs beyond 5 au and the expected number within that distance is small." Indeed, at another point in the paper, the authors indicate that an ISO would need to be "several hundred kilometers in diameter to be detected" and that the detection of an ISO of such a size within 50au of earth is "extremely improbable".[55] As a result, the authors tell us, "[t]he prospects for identifying a

[55] Oumuamua was detected at a distance of approximately .16175au and was only hundreds of meters large.

large chuck of material ejected by an extra-solar system passing through our own solar system appear to be bleak."

Oumuamua, then, if it is a natural object, should have never been seen. But when we consider Oumuamua's potential candidacy as a body of frozen molecular nitrogen, things only get worse. For starters, only 4% of potential ISOs could be of such a composition and the formation of such bodies has never been witnessed and is only theoretical. On top of that, the odds of a naturally occurring ISO appearing in the Local Standard of Rest—as Oumuamua did—are a probability of .2. Yet, when the frozen molecular nitrogen hypothesis was advanced, conventional science was so eager to accept it over Loeb's alien origins hypothesis that they completely ignored the statistical absurdities involved, and wouldn't even note them as a potential infirmity in their responses to Loeb.

A more recent argument in favor of a naturally occurring object is that Oumuamua was a hydrogen iceberg. While the proponents of this hypothesis have constructed a model showing how a hydrogen iceberg could, in theory, display propulsion characteristics similar to that displayed by Oumuamua, they do not have any direct or concrete information as to how such objects can form, nor any data on the likelihood or statistically frequency of such objects being formed or ejected from solar systems. Nor do the proponents of this view offer any explanation for Oumuamua's luminosity or its likely disc shape. While it is certainly possible that a hydrogen iceberg could mimic the non-gravitational acceleration of Oumuamua, the proponents of this theory offer no explanation or argument addressing the statistically implausibility of this theory, or how hydrogen icebergs may be formed of sufficient size to reach our solar system without being destroyed. Yet, this hypothesis is presented as if it has "closed the book" on the controversy surrounding Oumuamua.

The odds of Oumuamua being either molecular nitrogen, or being part of a hydrogen iceberg or some other natural object, are so bad that not even Han Solo would take them. Yet, the mainstream scientific community is so averse to accepting the possibility of Oumuamua being an alien anomaly, that they are willing to ignore the sheer statistical implausibility of Oumuamua being a natural object just to have the assurance of a more prosaic explanation. Moreover, they make no attempts to explain 'Oumuamua's strange luminosity or the 91% chance that it was a round disc shape.

Clearly, these features need to be accounted for before any hypothesis can make a claim to completeness. Yet, conventional scientists appear so desperate to keep alien anomalies out of our solar system, that they will simply ignore these glaring inconsistencies and failures of analysis.

4.10 *The scientific tribunal*

From a standpoint of scientific discourse, we find it highly problematic that proponents of a natural object theory for Oumuamua are never forced to account for the serious omissions in their reasoning. Even though neither the hydrogen iceberg or molecular nitrogen hypotheses are more plausible or complete from a scientific standpoint than the hypothesis advanced by Loeb, these alternatives are immediately embraced as definitive simply because they are more "normal" and more in accord with the paradigms under which mainstream astronomy already operates. Just as the SETI paradigm arbitrarily suppresses near-earth alien anomalies, or requires "extraordinary proof" for such claims, opponents of Loeb arbitrarily give greater weight to the "natural object" hypotheses of Oumuamua, even though their explanation is no more plausible than Loeb's. If the proponents of the natural object theory were resting their claims on objective scientific judgment alone (and were not being influenced by arbitrary factors) the best they would say is that their explanation is just as plausible as Loeb's.

The strange case of Oumuamua shows that scientific discourse, as constituted in its present form, is not able to coherently and rationally discuss alien anomalies. In the case of Oumuamua, we don't know exactly what the object is and probably never will. However, we do know that it is anomalous, and we do know that it does not fit the profile of any known "natural object". While we can construct hypothetical natural objects which it could be, it is also necessary to account for the actual presentation of Oumuamua to our field of observation, as well as its luminosity and disc-like shape. Yet, none of the proponents of the natural object theory ever address these aspects of the object in their logic, or are called to explain the aspects of Oumuamua that naturally suggest to the mind that the object was manufactured by an alien scientific process. Even though no one has demonstrated to a degree of reasonable scientific certainty that Loeb's explanation is any less plausible than any other, time and again, the "natural object" hypothesis is preferred and its substantial statistical infirmities routinely ignored.

For reasons that appear wholly arbitrary, mainstream scientific discourse will not even admit that Loeb's hypothesis for Oumuamua is as equally as plausible as any other, even though this statement is probably true. It appears that scientific discourse is able to evade the uncomfortable lacunae in the natural object theory due to the very nature of peer-reviewed scientific discourse itself, which imposes no constraints or regularities on scientific discourse. While the peer-reviewed model may work well in the "normal" sciences, it appears inadequate in the case of alien anomalies due to the strong current of what Jung referred to as the "numinosity" of the subject matter, which appears to have a tendency to override the application of basic scientific rationality.[56] In order to formulate coherent paradigms to situate these alien anomalies within a world of scientific meaning (regardless of whether these anomalies turn out to be real aliens or not), changes to the forms scientific discourse may be necessary to account for the strange powers that the subject of alien anomalies seem to exert on the human mind.

According to Kuhn, scientific paradigms are akin to "accepted judicial decisions in the common law" which are continuously articulated and specified "under new or more stringent conditions". Paradigms describe the methods and rules to be applied to various situations, and define the range of admissible evidence in a manner that is not entirely dissimilar to the ways rules of evidence are applied in a court of law.

---

[56] Jung describes this concept of numinosity in relation to alien anomalies as follows. "[B]esides its formal mode of manifestation the archetype possesses a numinous quality, a feeling-value that is highly effective in practice. One can be unconscious of this value, since it can be repressed artificially; but a repression has neurotic consequences because the repressed affect still exists and simply makes an outlet for itself elsewhere, in some unsuitable place. . . . the UFO comes from the unconscious background which always expressed itself in numinous ideas and images. It is these that give the strange phenomenon an interpretation that makes it appear in a significant light—significant not merely because they arouse dim historical memories which link up with the findings of comparative psychology, but because actual affective processes are at work." *Flying Saucers*, pg. 34.

> [T]he problems accessible within a given research tradition display something . . . like . . . puzzle characteristics. The man who builds an instrument to determine optical wave lengths must not be satisfied with a piece of equipment that merely attributes particular numbers to particular spectral lines. He is not just an explorer or a measurer. On the contrary, he must show, by analyzing his apparatus in terms of the established body of optical theory, that the numbers his instrument produces are the ones that enter theory as wave lengths. If some residual vagueness in the theory or some unanalyzed component of his apparatus prevents his completing that demonstration, his colleagues may well conclude that he has measured nothing at all. For example, the electron-scattering maxima that were later diagnosed as indices of electron wave length had no apparent significance when first observed and recorded. Before they became measures of anything, they had to be related to a theory that predicted the wave-like behavior of matter in motion. And even after that relation was pointed out, the apparatus had to be redesigned so that the experimental results could be correlated unequivocally with theory. Until those conditions had been satisfied, no problem had been solved.
>
> Similar sorts of restrictions bound the admissible solutions to theoretical problems.

Of course, scientific and judicial decisions differ in their methods of application, but at a very generalized level the kind of decision-making activity and analysis that occurs in the mind of a scientist when she is attempting to resolve an anomaly, is not all that different than the judge who is deciding how to apply the law to a very difficult set of facts and arrive at a final decision. Moreover, scientific discoveries are meaningless unless other scientists accept them as valid, acting in a sense as jurors do in a court of law. However, scientific discourse as currently constituted has no form of authority that guides and structures the discourse and scientists are free to believe whatever viewpoint suits them and avoid a direct examination of alien anomalies. While this model of discourse clearly works for the vast majority (if not all) of the normal scientific puzzles, it appears insufficient in the case of alien anomalies. When considering alien anomalies, scientific discourse is dissimilar to that of judicial discourse because a judge can't decide to ignore a case before her just because it may be difficult.

Because alien anomalies have strong emotional and subconscious effects on the minds of conventional scientists, they appear unable to think straight about these exotic objects. The reason for this lies in the fact that a variety of non-scientific and arbitrary factors influence the scientist's viewpoints on aliens and their ability to view alien anomalies in an objective manner. As we have seen in the case of Oumuamua, Loeb's hypothesis is no less plausible than any other, yet the natural object explanation is always accepted as more valid, even if the application of the natural object hypothesis requires wild leaps of logic, or belief in things that have never been seen or experienced by anyone. As we have seen in the analysis of the SETI paradigm, mainstream science has adopted an arbitrary attitude towards alien anomalies, choosing to keep them at a safe distance, instead of bringing them in for closer study. This tendency to instinctively avoid the difficult questions posed by alien anomalies appears to permeate the field of conventional scientific discourse. In order to countermand the effects of these arbitrary influences, a more structured form of scientific deliberation may be warranted in the case of alien anomalies; without that, it may be very difficult to create a positive momentum which might one day lead to viable paradigms of scientific knowledge.

Consider how these forms of judicial discourse might hypothetically apply in the strange case of Oumuamua. Let us assume that there exists a scientific tribunal of some sort composed of properly credentialled scientists from a wide range of views on the subject of alien anomalies to which an interested individual might submit a petition for a declaration regarding Oumuamua. Let us assume that the petitioner seeks a declaration that the "alien lightsail" hypothesis is as plausible as any other at this point in time. In submitting a request of this tribunal, the petitioner would provide findings and data regarding Oumuamua and request a specific decree. After submitting arguments and evidence, the petitioner could be questioned by members of the tribunal. The petitioner's proposed findings could be voted on, either rejected or accepted by the majority. Opponents of the petitioner could submit their own proposals and question the petitioner on her reasoning and analysis. Similarly, the petitioner could challenge her opponents as well. Through a deliberative process, a majority opinion of some sort could be rendered, and individual members of the tribunal would have the chance to offer concurrences or dissents addressing particular aspects of the case, to which other members could sign on. In this way, all the members of a tribunal would be able to share their viewpoints on the meaning of the evidence, and various theoretical implications of the study.

Under this type of proposal, coherent and structured discourse might emerge. Over time, these "decisions" regarding particular cases might serve as valuable precedent which members of the scientific community may find reliable or trustworthy when considering future alien anomalies. Of course, decisions of this nature do not "prove" anything, but simply develop a positive body of knowledge that has been credibly tested, examined, and considered in a structured, non-arbitrary manner. The passage of time may then allow a more robust body of knowledge regarding alien anomalies to emerge—something which cannot occur under the current peer-review model due to its lack of structure, and the ease with which conventional scientists are allowed to escape a thorough justification of their views. It may turn out that some cases are too inconclusive or speculative for anyone to arrive at firm conclusions; in other cases, results may be clearer. However, as a coherent body of knowledge is developed over time, scientists involved in this work may learn to see patterns and areas in which further observation, or increased or enhanced detection methods may be necessary. As a scientific "case law" of alien anomalies is developed, it may help to guide and direct avenues of further inquiry, possibly leading to the development of genuine "alien paradigms" which might lead to new forms of normal science.

This is simply one hypothetical way in which a structured discourse regarding alien anomalies could be approached, but there could certainly be other ways to do this as well. In *Structure*, Kuhn talks about the importance of the scientific process of "translation" as a means to rework the scientific discourse about paradigms in times of crisis. In this regard, the suggested use of the term "alien anomalies," as a valid substitution for any descriptions preceded by the terms "unknown" or "unidentified," is another way to work on a conversion process. Again, when scientists call an aerospace anomaly "unknown", or a flying object "unidentified", what they are really saying is that the anomaly *could have been caused* by an alien but they just don't know for sure because they have not been able to generate objective data regarding the occurrence as they were not prepared to observe it. Indeed, the only reason why they will not suggest a conventional cause for the phenomenon (i.e., "natural" or "human"), is because they can see that such a cause would not fit. However, the language they use to discuss such matters suppresses this truth from being spoken and discussed. This improperly inhibits the development of an actual paradigm regarding alien anomalies by making people feel like they are prohibited from directly discussing the ultimate causes of the anomalies and diverts their attention from discussing the real issue (i.e., their lack of preparedness to make a timely observation).

In order to countermand this inhibition, it may be necessary to use the term "alien anomaly" (or something similar) which simultaneously suggests that we don't know the cause for the anomaly, and that the cause could be an intelligent alien presence. Of course, saying that an anomaly "could be" caused by an alien is not equivalent to saying it "is" an alien. Figuring out whether something "is" an alien is the result of a separate scientific process that has not yet begun. We will not know for some time what these anomalies "really are", just like Priestly and Lavoisier did not "know" that they had discovered oxygen at the time that they had actually isolated it, like Roentgen did not "know" he had discovered X-rays until he locked himself in his laboratory for 7 weeks straight to figure it out, and like the peoples of the 17th century did not "know" the earth really moved in space until they saw things for themselves through Galileo's telescopes.

4.11 *Alien anomalies and the embrace of scientific process*

If the analogy between the Copernican Revolution and a revolution in alien anomalies is a valid one, then disclosure of alien anomalies is not going to be the result of a revelatory event by the specialized segments of the scientific community or other gatekeepers, but rather the result of a conversion process resulting from ongoing discovery and scientific inquiry. Also, we should probably brace ourselves for the possibility that the actual study of aliens might be as tedious as "patiently watching" ants, bees, or birds. The important thing is to start the process to see where it might lead. Only once the process has been in motion for some time may a paradigm emerge that might situate alien anomalies within a network of scientific discourse that will have meaning and lead to new scientific facts, products, and results—but, of course, there are no guarantees here and all this talk about aliens could be complete delusion. Indeed, an alien paradigm might never emerge, the anomalies will remain unresolvable, and we will be prompted to "close the book" on this subject and move on to something else more pressing. Either way, science should be able to provide some answers to these questions if given sufficient time and space to do its work; but until that happens, everything will remain "unknown" or "unidentified" to us and these alien anomalies will remain forever exiled to the realm of pre-scientific—but never realized—potential.

5.0 ALIEN PARADIGMS

5.1 Some alien anomalies bear the appearance of real phenomena

In 1947, the U.S. Air Force had determined that there were anomalous occurrences in the skies which were directly interfering with their ability to work. At the time, the Air Force was not looking to make scientific discoveries, but simply wanted to make sure that the airspace over the United States was adequately protected. There was no theoretical axe to grind here; instead, the Air Force simply wanted to know what was going on and how it should be dealt with. This was a genuine *need to know* which arose under circumstances which strongly indicate that something real was going on and that the pilots seeing these things were looking for real answers.

Because the Air Force wanted to gain actual knowledge, it turned to a valid scientific approach to alien anomalies as reflected below in the following passage of the declassified[57] Project Sign report.

> This report is written to present the status of work being accomplished on Project "Sign", to summarize the data collected on sightings of unidentified aerial objects, to review the methods and reasoning applied in the evaluation of the data, and to present the results so far obtained from the study of data available.

This is the kind of approach that trained professionals would employ. It is difficult to credibility argue that they were taking these steps for no good reason.

Clearly the case of the California tic tac also stands out as another alien encounter that is difficult to dismiss. In our discussion of it, we just cited to one report and analysis of the incident, but the reader who doubts that it really happened, and doubts that the tic tac encounter was just as real as any other human experience or occurrence—and self-evidently not the product of human error or

---

[57] The fact that the report was classified initially is also a strong factor indicating that this wasn't viewed as just a joke or delusion, but rather as a real problem that had to be dealt with in the real world.

delusion—is encouraged to do their own research, as they say. If you don't believe the tic tac encounter happened, watch the videos, listen to the innumerable interviews given by the pilots involved, assess their credibility, consistency and motives, and examine whether any plausible explanation has ever been advanced to explain what these pilots saw. The totality of the circumstances makes it impossible to rule out the possibility that the tic tac encounter was caused by a real alien being in earth's aerospace as there truly is no conventional explanation for what was experienced and seen. *See* **Section 7.1** ("working out an alien paradigm").

At the same time, since we simply do not have the kind of surveillance and detection systems in place to adequately detect and measure physical proof of these occurrences, and others like them, we are not in a position to say what they are. We can say, however, that some of these incidents bear strong indicia that they did occur. We can also say that these types of occurrences sometimes seem just as real to those who experience them as any other type of experience in their lives, even if the experience itself is fundamentally alien in nature.

5.2 *Approaching alien anomalies as pre-scientific phenomena*

When a person considers the topic of alien anomalies in a serious way, there usually comes a point in time where that person decides if they want to "believe" or not. But there is another way to look at this situation. Instead of feeling compelled to make a determination as to whether certain alien anomalies are real or not, one can simply say that, regardless of whether they are real, they are worth investigating further. The question then becomes what form of investigation is worth pursuing.

There are different avenues here, of course. The most common form of investigation into alien anomalies at present consists of studying anomalies that have already passed and examining the forms of direct and circumstantial evidence the anomaly has left in its wake. The value of such an inquiry is limited by the forms of evidence that have been gathered, however. If the anomaly occurred in a situation where no controlled detection was established in advance, then the value of the evidence is largely anecdotal and non-rigorous.

A related form of inquiry is to study and catalogue the alien experiences that people have had during their encounters. Oftentimes, persons who claim to have had such experiences will testify as to how the experience affected them and how it changed their view of the world. These reports have value primarily of philosophical, religious, or metaphysical nature and are anecdotal. *See **Section 6.0*** ("alien experiences").

A different approach is to place alien anomalies in the category of pre-scientific phenomena, similar to the state occupied by Bacon's studies referenced by Kuhn as discussed in **Section 4.1.** This approach acknowledges that alien anomalies have some weight and value, but also takes the position that they have not been observed, measured, and studied over a sufficient length of time, and with a sufficient level of detail, to render them into what we might refer to as "normal" scientific objects. Anecdotal evidence is not disregarded but is instead taken for what it is; a step along the way to a different form of knowledge which has not yet ripened into being.

5.2.1 *The ether and the atom*

Getting someone to believe something happened is not always that big of a deal. People in the real world end up believing stupid shit for terrible reasons all the time. This is not to say that belief is worthless or silly; especially in the realm of scientific inquiry. For without belief, or "scientific faith", so many scientific objects would never have come into being in the first place since they were not directly detectable until sufficient effort had been expended for a sufficient period of time. Without the power of that belief animating and propelling a given scientist to continue on with their work, many important discoveries would have never been made.

We have previously referred to the ether and the atom in this work. Both were believed to be real at one point in time. We no longer believe the ether is real, but we "know" the atom is real. Why? What's the real difference between these objects today?

Let us recall that there were very sane and rational reasons for "believing" that both were real at one point in time. For the atom, the idea of basic building blocks that can be combined in certain orders to make larger objects in the human world makes a lot of sense. For the ether, as we noted previously, the wavelike property of light naturally suggested a medium like the ether.

The atom was eventually "discovered" but the ether was not. However, the atom is not what it seemed to be at the time of discovery, for we later learned there was a lot more going on in the space of the atom than was first suspected. But the atom is at least a genuine *intermediary object* in the quest to know more about the structure of reality and work with it directly. It is a real entry point into different forms of knowledge and technology.

By contrast, the ether is entirely fictious today. However, before the notion of the ether was dispelled, people reasonably theorized that there had to be something like a medium for light to explain its wavelike properties. At that time, the paradigm known as "relativity" did not exist until the work of Einstein. Once the Einsteinian paradigm was articulated, the search for the ether was deemed useless. Instead, a new and hitherto unknown domain of knowledge was revealed.

What, then, is the real difference between these two cases from the standpoint of human knowledge? All we can really say is this: people took a belief about the nature of certain phenomena in the world, thought of ways to test, record, measure, and examine it, conducted those tests, examinations, and measurements, developing new techniques and devices when necessary to do so, and, in that process, patiently waited for the world to give them results. These results then became new facts about the human world and changed it in certain ways forever.

5.2.2 *Mendeleev's "vision" of the periodic table of elements*

Mendeleev's vision of the periodic table of elements is a most excellent example of modern-day paradigm development and conversion. This true story has all of the elements needed to deeply consider how real alien paradigms come into formation and change the human world.

As detailed in Paul Strathern's book, *Mendeleyev's Dream: The Quest for the Elements* (2019), initial chemical observations were likely made by the 16th visionary Paracelsus who was the first person to describe the properties of what we today would term "elements", in his differentiation of bismuth and cobalt. "Paracelsus first encountered bismuth in the Fugger mines" of Austria, Strathern tells us.

> The Austrian miners of the period were firm believers in the Aristotelian theory of metallic evolution. According to this, the long process of natural transmutation resulted in three distinct types of lead: ordinary lead, tin, and bismuth. The latter was closest to silver. As a result, when they uncovered a new vein of bismuth, the miners were in the habit of exclaiming: 'Alas, we have come to soon.' In other few years, they believed, it would have become silver.
>
> Paracelsus described the properties of bismuth, but without of course realizing it was an element. He simply found he was unable to break down this substance any further with the techniques of his disposal. Chemistry was not yet a coherent science, more a growing proliferation of techniques. But it was becoming evident that something was emerging from the alchemical hell's kitchens: a subject that was distinctly separate from the quest for gold. It has been claimed that Paracelsus was the first to refer to this subject as chemistry.

Paracelsus discovered bismuth as an intermediary object whose status was not defined until the revelation of the periodic table of elements many centuries later. But still it was an object that had been isolated from other objects and could be studied in its own right. We can say that Paracelsus discovered "bismuth", but it also wasn't the same bismuth that we know today, just like the "oxygen" of the 18th century is not the same oxygen of today. The physical object itself may have been the same, but its reality in the human world was far different. At the time Paracelsus discovered bismuth, it was a pre-scientific phenomena, located somewhere between the mists of alchemy and the realm of Aristotelian logic.

Other examples of this type of pre-scientific discovery abound in the historical record with regards to chemistry. These are cases where something real is discovered, but what we today would consider the real and true understanding of the object is not known for a long time after the object itself is initially discovered. At the time that bismuth was discovered, for example, it was somewhat of a fantasy object in the sense that people of that time truly believed it would have later turned into silver if it had been left alone (a view which we today seems to us as childish and fantastical). During the period of time when the object is not understood in the way we understand it, it is effectively a different kind of object for a different paradigm of understanding. To us today, the understanding appears incomplete and we see the errors quite clearly. Still, the existence of error does not inhibit the development of knowledge as something real has still been discovered.

Another example of this kind of thing is seen in the example of the 16th century Flemish physician Jan Baptista van Helmont. Strathern describes Van Helmont's "most memorable experiment" as follows.

> [He] placed 200 lb of dried earth in a large earthenware vessel; this was then watered and planted with a willow shoot weighing precisely 5 lbs. The vessel was covered to protect it from accumulations of dust, and daily watering continued, using only distilled water . . . After five years of watering, the willow had grown into a considerable tree, which van Helmont then dug up and weighed. He found that the tree weighed 169 lb 30 oz. He then dried out the remaining earth, weighed it, and found that it had lost just 2 oz . . . Van Helmont believed that his experiment proved that everything was made of water . . . [it] is widely regarded as the first application of measurement to an experiment involving both chemistry and biology, making the inception of biochemistry.

Van Helmont clearly learned something real about biological growth processes here, but what he took away from it (mainly that "everything is made of water") does not look like a scientific conclusion to us; instead, it looks like an erroneous Aristotelian or fantasy conclusion. But still, he learned something about biochemical processes. He saw there was something real going on in the world which could not be directly seen with unaided human perception and was in fact invisible until a specific observational regimen was imposed upon the process. Again, we see the importance that a stable period of observation, under controlled conditions, plays in the development of genuine knowledge. These are the kinds of stable observational conditions that should be created around alien anomalies in order to study them, taking them out of the realm of fantasy speculation and mystery and bringing them into the clear light of patient observation. Even if all the details are not comprehended, something can still be learned.

Another example of this form of pre-scientific experiment is the that of the 17th century physicist Evangelista Torricelli to detect the existence of air pressure, which Strathern describes as follows,

> . . . he took a long glass test tube and filled it with mercury. After placing his thumb over the open end, he inverted the tube and placed it in a vessel also filled with mercury. He found that the mercury in the tube sank to a certain level. This was always approximately 76 cm higher than the surface of the mercury outside the tube, which remained open to the air. Torricelli concluded that it must be the weight of the outside air that forced the mercury inside the tube to a higher level than the mercury outside. Torricelli had discovered air pressure.

After conducting further variations on this experiment and socializing them with others (who also conducted the same type of experiments in different places and under different conditions), the contours and dimensions of air pressure as an intermediary object were established. "Thus, three centuries before the first rockets were fired into space," Strathern tells us, "it was known that the earth's atmosphere extended only a certain finite distance from its surface. Arguably, Torricelli's experiment had led to the discovery of space itself."

Toricelli's experiment, while still pre-scientific because it lacked the comprehensive paradigms which would allow it to be understood in the way that we see it today, established the real existence of a phenomenon which was ever present, but invisible and undetected until that time (i.e., something called "air pressure"). But what air pressure meant to Toricelli would have been far different than what it means to us today. Again, the paradigm may be incomplete and filled with error as we look back from today's vantage point, but it still led to the distillation of real knowledge.

Another example of pre-scientific investigation is the scientific spectacle performed by the 17th century German engineer and inventor, Otto von Guericke, on May 8, 1654, demonstrating the existence of the vacuum, which Strathern describes as follows,

Guericke's most famous experiment was carried out on 8 May 1654 before the Emperor Ferdinand III, and attracted vast crowds from all over Saxony. This time the experiment had two large hollow copper hemispheres which had been precisely cast so that their rims fitted tightly into one another. (These were later to become known as the Magdeburg Hemispheres.) The Emperor sat enthroned on his platform, above the ground gathered in the sunlit square before the parliament building. All watched expectantly as Guericke greased the rims of the two hemispheres and carefully fitted them together. Then the blacksmith began vigorously pumping the air from inside the sealed copper globe. After a while he was joined by his assistants, as the cranking of the pump gradually became more and more laborious. The crowd then watched in bemusement as a team of eight horses harnessed together was led into the square and attached to one hemisphere of the copper globe. Another team of eight horses was then attached to the other hemisphere. At a signal from Guericke, the two teams of horses strained forward in opposite directions, attempting to pull the two hemispheres apart. The crowd fell silent as the powerful drayhorses heaved, but no matter how they were whipped they were unable to separate the hemispheres. Guericke now addressed the Emperor and the crowd. This was not a trick, he told them. All that was holding together the two hemispheres was the pressure of air surrounding them. The vacuum inside the globe meant there was no opposing pressure to balance this great outer force—which was even more powerful than the strength of sixteen horses. The Emperor was amazed, an emotion that was clearly echoed on the gaping faces of his gathered subjects, who now began to applaud. But Geuricke held up his hand, silencing the crowd. The experiment was not over yet. The horses were untied from their harnesses and led away. Guericke then fiddled with the pump. The crowd craned forward curiously. There was a sudden hissing sound as the outside air pressure rushed into the hollow globe to fill the vacuum. Then, without warning, the two copper hemispheres simply fell apart of their own accord. Now that there was no vacuum, and the pressure inside was the same as the pressure outside, there was nothing to hold them together.

> Guericke's experiment soon became so famous that he was demonstrating it all over Germany. Tales of his feat with the Madgeburg hemispheres spread across Europe; etchings were made of it; and when a garbled version crossed the Channel to England, it was said to have given rise to the nursery rhyme about Humpty Dumpty.[58]

As a result of these types of experiments and others, Strathern argues that "the unknown properties of matter. . . were at last beginning to emerge . . . The field of chemistry was now ripe for the intervention of a great scientific mind. This duly appeared in the form of Robert Boyle." Boyle's work can be summarized as follows. Born in 1627, as a young man under the tutelage of a private tutor, possessed by religious fervor, a sense of divine purpose, and a degree of melancholic gloom, he absorbed the work of Galileo and the mechanistic paradigms of Descartes which, according to Strathern would not have been available at almost any university in Europe. In other words, Boyle was absorbing paradigms which were not part of the convention corpus of thought at the time. "Even at this late stage," Strathern writes, "Aristotelianism remained the academic order of the day. Boyle's lack of orthodox formal education was to prove the making of him." Boyle, we might say, was exposed to alien paradigms in his youth.

Boyle decamped in Oxford, England in 1656, and immersed himself in the experimentalist undertakings of Francis Bacon. He became a member of what was called, "The Invisible College", later becoming the Royal Society in 1662. Their motto was *'nullius in verba'*, which roughly translates as "nothing by word alone".

This unconventional mindset freed Boyle to experiment in ways that were not constrained by the dominant paradigms of his day. Boyle took the work of Galileo and Guericke and quantified it in greater detail, proving "Galileo's prediction that in a vacuum any two bodies would fall at precisely the same rate of speed", confirming that sound did not travel in a vacuum, and making various preliminary findings regarding the conduction of electricity and air. Among other things, Boyle "proved once and for all that air was no mystical entity which somehow pervaded the entire

---

[58] The fact that Guericke's experiment morphed into a nursery rhyme is a fun example showing how anomalies in the real world can so easily turn into rumors.

world. Air was not an essential element of nature as the Aristotelians believed; it was simply a substance with definite properties of its own." Boyle also devised ways to "separate gases . . . and study them as entities entirely separate from air. He noticed that they all possessed this property of elasticity." Following a series of experiments, he articulated the famous "Boyle's Law, which states that the volume of gas varies inversely according to its pressure." Boyle's Law still holds good today, which simply means that we consider it properly scientific. Again, we see how a precise form of study draws an anomaly from the vapors of mysticism through precise observation which reveal new features that can be quantified, observed, and studied. The object then becomes workable, instead of being merely an object of fantastical speculation.

In 1661, Boyle wrote *The Sceptical Chymist*. Strathern contends that this book provided the foundation of modern chemistry. The *Sceptical Chemist* contained accounts of experiments and their results, written in "a clear and easily comprehensible manner," allowing them to be replicated by others. "From a secretive practice in a murky den," Strathern writes, "the new chemistry was transformed into a universal science which could be practiced in any laboratory anywhere." The work defined chemical elements, in a strong break with the Aristotelian framework, as "certain primitive and simple, or perfectly unmingled bodies; which not being made of any other bodies, or of one another, are the ingredients of which all those called perfectly mixed bodies are immediately compounded, and into which they are ultimately resolved." Boyle also proposed that the elements could combine together and form compounds, and concluded that "all these compound substances depended for their properties upon the number and position of the elements they contained." We can see in this formulation a statement of a paradigm which resembles what we would consider to be modern chemical science.

In addition to articulating new paradigms for the science of chemistry, Boyle continued on the path of active experimentation as well. Strathern says that Boyle's greatest experimental contribution to the science of chemistry was "his discovery of a method for distinguishing between acids and alkalis. He found that syrup of violets turned red with acids, green with alkalis, and remained the same with neutral solutions. At last, a proper definition of acids and alkalis had been discovered."

Eight years after Boyle defined the term "element" in *The Sceptical Chemist*, the German scientist Hennig Brand discovered the first new element by isolating phosphorous. The method he used to isolate this element is described by Strathern as follows,

> Brand began a prolonged and extensive investigation of the properties of human urine, which must have sorely tried the patience of his well-to-do wife, to say nothing of his neighbors. He collected fifty buckets of human urine, which he then allowed to evaporate and putrefy until they 'bred worms'. This he then boiled until there was a pasty residue. When he left this in the cellar for some months he found it had fermented and turned black. Doubtless with the entire neighborhood now up in arms, Brand proceeded to heat the black fermented urine concentrate with double its weight of sand in a retort, whose long neck was plunged into a beaker containing water. The final distillate collected under water in the beaker was a transparent waxy substance. When removed from the water it glowed in the dark, and sometimes even spontaneously ignited, giving off white dense fumes. He decided to name this new substance phosphorus, from the Greek *phos* ('light') and *phoros* ('bringing').

Once again, we see the mode of pre-scientific experimentation at work. Obviously, Brand did not know exactly what he was going to find, and even though he named his discovery "phosphorus", this was not the same phosphorus that we know today. Today's phosphorous resides within a network of meaning that Brand couldn't have even imagined. For Brand, the real paradigm that would have explained what he discovered was alien as hell at the time he made his discovery; to us, however, it is simply the way things are today. Obviously, Brand was groping in the dark, and who knows what kind of insight and hunch prompted him to engage in the methodological inquiry that he did. This example again shows the kind of pre-scientific activities that one must engage in order to learn real things about the world.

Brand refused to publicly release the method he used to distill phosphorus for some time but eventually sold it to a certain Dr. Johann Krafft of Dresden, who then gave phosphorous demonstrations all over courts in Europe. Boyle was able to view one of the demonstrations and was able to figure out how to prepare phosphorus on his own. Boyle then wrote out the method to prepare phosphorus in clear, step-by-step instructions which he provided under seal in a letter to the Royal Society. Eventually other methods of producing phosphorus were discovered.

According to Strathern, Boyle's approach opened the door for other experimenters over time. In 1770, Scheele discovered chlorine, barium, and molybdenum. As noted previously, Scheele also played a critical role in the discovery of oxygen. Others in this same time period discovered the elements of nickel and platinum.

As the classification process continued, there was an effort to develop a paradigm that explained the laws giving rise to these various elements, as well as gases. This eventually led the English scientist Dalton to derive his atomic theory. Kuhn describes Dalton's process in *Structure* as follows,

> Dalton was neither a chemist nor interested in chemistry. Instead, he was a meteorologist investigating the, for him, physical problem of the absorption of gases by water and of water by the atmosphere. Partly because his training was in a different specialty, and partly because of his own work in that specialty, he approached these problems with a paradigm different from that of contemporary chemists.[59] In particular he viewed the mixture of gases or the absorption of gas in water as a physical process, one in which forces of affinity played no part. To him, therefore, the observed homogeneity of solutions was a problem, but one which he thought he could solve if he could determine the relative sizes and weights of the various atomic particles in his experimental mixtures. It was to determine these sizes and weights that Dalton finally turned to chemistry, supposing from the start that, in the restricted range of reactions that he took to be chemical, atoms could only combine one-to-one or in some other simply whole-number ratio . . . For Dalton, any reaction in which the ingredients did not enter into fixed proportions was *ipso facto*

[59] According to Kuhn, at that time chemistry was dominated by a theory of chemical affinity which stated that "the elementary atoms of which all chemical species consisted were held together by forces of mutual affinity. Thus, a lump of silver cohered because of the forces of affinity between silver corpuscles (until after Lavoisier these corpuscles were themselves thought of as compounded from still more elementary particles)." Today, the theory of chemical affinity appears to us as a fantasy idea.

> not a purely chemical process. A law that experiment could not have established before Dalton's work, became, once that work was accepted, a constitutive principle that no single set of chemical measurements could have upset . . . What chemists took from Dalton was not new experimental laws but a new way of practicing chemistry (he himself called it the "new system of chemical philosophy"), and this proved so rapidly fruitful that only a few of the older chemists in France and Britian were able to resist it.

In the wake of Dalton's work of discovery for precise ratios of atomic weights and the relationship between those weights and the characteristics of the element themselves, scientists operating in this domain continued making various attempts at listing and categorizing the elements in a more systematic way. In this vein, the Swiss chemist Jons Berzelius introduced the modern chemical notation, using various letters and subscripts, and also introduced the idea that various elements had negative and positive charges. This new language allowed the relations among the various elements and their weights to find a direct and workable expression. But still, an overall paradigm which organized and explained the workings of the chemical world remained a mystery.

It is in this context that Mendeleev uncovered the Periodic Table of Elements. By early 1869, Mendeleev had completed the first volume of a projected two-volume work entitled *Principles of Chemistry*. At that time, Mendeleev was 32 and a professor of general chemistry at the University of St. Petersburg. He had studied with the best practitioners in field of chemistry throughout Europe. He had already published important papers in the field of chemistry, and written the first chemistry textbook for Russian students, a five-hundred page book that he had produced in just sixty days. His knowledge of the field of chemistry was prodigious. His lectures at the University of St. Petersburg were well attended. As described by one student, "The hall was always crowded with something like two hundred students, many of whom, I'm afraid, could not follow Mendeleyev, but for the few of us who could it was a stimulant to the intellect and a lesson in scientific thinking which must have left deep traces in their development, as it did in mine."

According to Strathern, Mendeleyev was well aware that chemistry was a vexing and imprecise science at time, and it was his goal to bring order to the field. As stated by Strathern,

> Mendeleyev was highly conscious of this defect in his chosen subject. His chemistry lectures may have included miscellaneous asides on everything under the sun, but his guiding intellectual principle was synthesis. His mind was constitutionally averse to leaving a fact to stand by itself. As he put it: "The edifice of science requires not only material, but also a plan, and necessitates the work of preparing the materials, putting them together, working out the plans and symmetrical proportions of the various parts. For Mendeleyev this was more than just an intellectual task: "To conceive, understand, and grasp the whole symmetry of the scientific edifice, including its unfinished portions, is equivalent to tasting that enjoyment only conveyed by the highest forms of beauty and truth."

The first volume of *Principles*, which Mendeleyev had finished, covered the halogen group of elements, all of which combined easily with potassium. The second volume was to focus on the alkali metals, which contained sodium and potassium. Mendeleyev was planning to have the first two chapters of Volume II cover this material.

On February 14, 1869, Mendeleyev was working on the problem of what group of elements he wanted to tackle next and how the elements would be ordered. He was under a certain time pressure, as he was scheduled to leave St. Petersburg to deliver a speech to a delegation of cheese makers and conduct a three-day inspection of some farms. Strathern describes his mindset at this juncture in the following terms,

> Painstakingly, Mendeleyev began sifting through his encyclopedic knowledge of the chemical elements, in search of some pattern of properties which might link groups of similar elements. There had to be a key to it all somewhere. The elements couldn't just have a random set of properties; that would be unscientific. Admittedly, they had a definite order of atomic weights, but that surely was only part of it. That just accounted for physical properties. What about an order amongst the chemical properties? . . . Again and again he racked his brains. Again and again he tried building up the elements one by one to form some kind of structure. But each time the structure tumbled to the ground like a pack of cards.

> By the morning of Monday 17 February Mendeleyev still hadn't come up with anything.

Mendeleyev didn't have much time before he had to catch the train out of St. Petersburg and was feeling pressured to find a solution before he left (Strathern says that Mendeleyev was conscious of the fact that he would become very depressed when he failed to solve a puzzle that he knew he was close to solving). On that morning, Mendeleyev began listing out the elements in order of their atomic weights and playing with various arrangements. But he kept coming up short. Then he tried something different.

Apparently, Mendeleyev was a fan of the card game patience. In patience, cards are aligned according to suit and descending numerical order. Mendeleyev began writing the properties of elements on blank cards and playing with their arrangements. But still nothing was clicking. Allegedly, he told a friend who had visited him during this time that "It's all formed in my head, but I can't express it." After continuing at this frustrating process for some more time without success, Strathern relates that the following occurred.

> By now the fading dusk outside the window had given way to the frigid darkness of night. Despite his growing exhaustion, he knew he couldn't stop now. Mendeleyev was possessed by the feeling that he was on the brink of a momentous discovery.
>
> So it comes as something as a disappointment to record that at that point he was overcome with fatigue. He leaned forward, resting his head on his arms amidst the scattered cards on his desktop. Almost immediately he fell asleep, and had a dream.

Mendeleyev later described the visionary state he went into as follows,

> I saw in a dream a table where all the elements fell into place as required. Awakening, I immediately wrote it down on a piece of paper.

This "dream" revealed the Periodic Table of Elements. The table generated by Mendeleyev listed the elements in order of ascending atomic weights from top to bottom and then created horizontal rows which grouped elements with similar graduating properties. The Periodic Table displayed a pattern of elements and had places for each of elements. The placements were not random but based on the weights and various properties of the elements.

The initial version of the Periodic Table was different than what one would see today in a modern chemistry classroom, as it has been refined and added to over the years. But the basic ordering idea behind the Periodic Table has remain unchanged. The pattern for the initial version of the Periodic Table contained placements and an arrangement for nearly all of the known elements based on their various properties, and in that regard made good sense. But it also contained a number of anomalies that called its correctness into question. How Mendeleyev dealt with and resolved these anomalies shows that he understood the paradigm expressed in Periodic Table as an independent existing entity that displayed real forms of knowledge.

For example, there were some elements in the horizontal columns whose weight did not conform to the order they were in. Mendeleyev responded to that anomaly by claiming that the experimental weight of the element was in error, and that the weight predicted by the table was correct. Furthermore, there were "gaps" in some of in some of the horizontal rows. Mendeleyev claimed that the gaps represented elements that had not yet been discovered. The Table predicted both the specific weight of these elements and the specific types of properties they would have.

One example of such a missing element was the hypothetical element that Mendeleyev named "eka-aluminum". This element, per the instructions from the Periodic Table, would have an atomic weight of 68, and properties between those of aluminum (whose weight was known to be 27.1) and uranium (whose weight was known to be 116). No such element had yet been discovered when the Table came into being.

Mendeleyev published his work on March 1, 1869. In the late summer of 1874, a French chemist named Lecoq claimed to have discovered a new element in a sample of zinc sulfide and named it gallium. Strathern describes what happened next as follows.

> Lecoq's newly discovered element was found to have an atomic weight of 69, and its properties indicated that it belonged to the boron group, between aluminium and uranium. The new element gallium all but exactly matched the properties which Mendeleyev had predicted for eka-aluminium. But when Lecoq calculated the specific gravity of gallium, he found it to be 4.7—whereas Mendeleyev had predicted that eka-aluminium would have a specific gravity of 5.9. Here was a glaring discrepancy which could not be overlooked. Was it possible that Mendeleyev's other "predictions" had been nothing more than a series of lucky guesses?
>
> As soon as Mendeleyev heard that Lecoq's finding did not match with his theoretical prediction, he reacted in characteristic fashion. Immediately, he dispatched a letter to Lecoq informing him that his sample of gallium was insufficiently pure, suggesting he repeat the experiment with another sample. Lecoq dutifully repeated his experiment with a larger sample, which he subjected to rigorous purification. And this time he found that the specific gravity of gallium was 5.9, just as Mendeleyev had predicted!

The similar thing happened 5 years later with the element germanium. This element appeared in a gap in the Table that Mendeleyev called "eka-silicon". It was supposed to have an atomic weight close to 72, its specific gravity would be 5.5, its oxide would have a specific gravity of 4.7, and its compound with chlorine would have a specific gravity of 1.9. Germanium showed up in exactly the place Mendeleyev said it would and had all of these precise properties.

Today, of course, the "correctness" of the Periodic Table is well-established for us. But when we look at the time period surrounding its initial formation in the normal world of people, we can see that the knowledge it displayed conflicted with other accepted knowledge at the time. Yet, Mendeleyev's belief in his own process, and the results he arrived at, overpowered the contrary evidence and the anomalies. Mendeleyev refused to bow to conventional systems of knowledge.

But what would have happened to Mendeleyev's Periodic Table if he discounted it because it conflicted with other knowledge that was the result of real life measurement (such as the initial properties of the newly discovered gallium), or because it contained anomalies instead of being complete?

At right about the same time that Mendeleyev made his discovery of the Periodic Table, a German chemist named Julius Meyer also made nearly the same discovery, but never got the credit that Mendeleyev did. One reason for this was because Meyer waited to publish his paper on the subject almost a year after his discovery, and Mendeleyev published within a few weeks of the revelation. But another reason why Mendeleyev got credit for his work, and Meyer did not, was explained by Strathern as follows.

> [D]ecisively, Meyer's conclusions were more tentative. He couldn't fully account for the anomalies in his table—the elements which were out of order, those which didn't appear to fit into their apparent group and the glaring gaps. When critics pointed out the discrepancies between Meyer's 'law' and the facts, Meyer had no defence. Mendeleyev, on the other hand, took the offensive. He was willing to back his chemical insight in the face of all the "facts".

Although the Periodic Table of Elements is one of the keystones of modern-day scientific conventional thought, at the time it emerged, it was most certainly an "alien paradigm". It was not, of course, a paradigm about aliens, but rather alien in the sense that it is unfamiliar, strange, suspicious, or simply "other". *See* ***Section 2.12.*** Not only did the Periodic Table a constitute a radically different paradigm with respect to the field of chemistry, but it also conflicted with conventional knowledge and was internally anomalous. It could have easily been rejected from the conventional viewpoint.

And although there had been prior attempts made to categorize and classify the elements prior to Mendeleyev, the Periodic Table was not simply a classification of existing knowledge but provided entirely new forms of knowledge solely from the intuition that Mendeleyev had in his visionary state. We truly cannot overstate the strange predictive power that it had. Prior to the Table, chemists had no way to tell what kinds of elements they might discover. But the Periodic Table predicted with stunning precision the precise details of unknown elements. Prior to the arrival of the Periodic Table, it would have been insane to even dream of a paradigm with such precise predictive power. While we can certainly say that all the puzzle pieces to put together the Periodic Table were at hand, the precise nature of the Table, and the power that it possessed, was completely alien in relation to the field of chemistry in the late 19th century. This wasn't just an interesting ordering of elements, but something much more than that—it was the birth of a new order of knowledge.

For us today, it may be difficult to really feel just how alien Mendeleyev's work was in its time. We cannot, in fact, imagine a world where there was no Periodic Table and where the understanding of chemical elements was still emerging from a "pre-scientific" to a status that we today would call "scientific".[60] We cannot turn back the clock and suddenly forget everything we know and experience what that kind of world was like.

One can use the lessons gleaned from Mendeleyev's approach in thinking about future paradigms that might emerge which seek to describe actual "alien anomalies". At present, evidence relating to alien anomalies appears far less developed than evidence related to chemistry was at the time of Mendeleyev, but one could envision a time in the not-too-distant future when the direct evidence of alien anomalies becomes more developed but still generally incomprehensible in a way similar to that of chemistry prior to Mendeleyev.

5.2.3 *Alchemy as an alien paradigm*

By the time the discipline of chemistry reached Mendeleyev, it had been purged from alchemical origins. After Boyle, this discipline looked much more like the chemistry we know today, than the alchemical investigations of the past, but was still not what we would consider chemistry. Today, the Periodic Table articulates an

---

[60] There is a certain degree of relativity between and among various paradigms here which we should be conscious of. No matter how correct, stable, and real the Periodic Table feels to us today, our knowledge of the history and sequence of paradigm development since Copernicus should make us leery of assuming that the Periodic Table is completely correct. Nevertheless, when we look back at Mendeleyev's work from our present vantage point, it still appears correct to us. But can we rule out the possibility that a future group of humans whose views and attitudes are unknown to us today, will always see the Periodic Table in the same way we do? It seems difficult to make such an assumption.

era of thought about chemistry from which we have not yet emerged, but we have no good reason to believe that all future generations will be bound by the Periodic Table; there could be future forms of knowledge that would expose errors and limitations in the Periodic Table that we cannot even imagine.

The conventional take on this situation is that alchemy was nothing more than an irrational disaster that prohibited real scientific discovery for centuries. But if we view alchemy as just another alien paradigm, we see how it served as a necessary—but not sufficient—cause in the development of chemistry proper.

We can briefly summarize the metaphysical and alchemical origins of chemistry up to the time of Boyle (which marks the beginning of paradigm that found its relative completion with Mendeleyev) as follows. According to Strathern, "[t]he earliest recognizable chemists were women, the perfume-makers of Babylon, who used the earliest known stills to produce their wares. The first individual chemist known to history was 'Tapputi, the perfume-maker,' who was mentioned on a cuneiform tablet from the second millennium BC in Mesopotamia." The Phoenicians used lead in their anchors. Tin was extracted from Cornish mines in England. Bronze was synthesized from tin and copper in the Mediterranean region around 3000 B.C. Mercury was known to the ancient Chinese and Hindu cultures, and found in Egyptian tombs dating back to 1500 B.C.

What we can say, generally speaking, about the origins and development of modern chemistry is that it began with physical practices working with matter at the level of chemical elements and compounds, but that the paradigms guiding those practices were badly out of alignment, or had no alignment at all, with what we today would call the "correct" chemical paradigm (i.e., The Periodic Table). The people of the past were working with matter in a way that had meaning in their lives, but when we look back from the standpoint of our paradigms, we see they lacked understanding as to what they were really doing. We also see how their lack of understanding limited their ability to work with chemical elements.

After Greek philosophers such as Anaximenes, Leuccipus, and Democritus, articulated paradigms that questioned whether or not matter was continuous, or composed of individual units ("atoms") which could be observed and/or manipulated, a new approach to chemical practices was opened up as there was a new paradigm through which to view them. Strathern says that this line of thought converged with a much older form of learning, known as the Egyptian art, or *khemeia*, in Alexandria around the 4th Century B.C. Strathern describes *khemeia* as follows,

> The origins of *khemeia* are lost in time. The word occurs in various ancient Egyptian hieroglyphs in connection with the burial of the dead. The Roman historian Pliny even says that Egypt itself was originally called *khemeia*, or black, after the dark rich soil of the Nile delta. The knowledge associated with this dark art is mentioned in several early sources, including the Book of Enoch (one of apocryphal books of the Old Testament). According to this source the secret knowledge of *khemeia* was passed on to a number of women by fallen angels in order to win their favor.
>
> Initially, this knowledge consisted largely of the chemical processes involved in embalming the dead . . . However, its practices soon evolved to include other chemical processes which had been discovered by the ancient Egyptians such as glass-making, dyeing, and, especially, the art of metallurgy. Thus, *khemeia* became connected with the seven known metallic elements: gold, silver, copper, iron, tin, lead, and mercury.

Strathern says that the "earliest known adept of the dark arts was a certain Bolos of Mendes, a Hellenized Egyptian who lived around 200 B.C. In his principal work later called *Physica et Mystica* ('The Physical and the Mystical') he listed a wide variety of esoteric experiments each of which ended with the incantation: 'One nature rejoices in another. One nature destroys another. One nature masters another.' This chant contains a tantalizing hint of true chemical understanding." When Strathern refers to this tantalizing hint, what he means is that we see in this ancient approach the seeds of our modern chemical knowledge. Looking back, we know that the persistent approach to breaking compounds down, looking for invisible undisclosed processes, and understanding how those processes could be used by us, was precisely the type of activity that made it possible for Mendeleyev to one day emerge. However, it would be impossible for Bolos to have any paradigm of knowledge even remotely resembling the Periodic Table.

From the time of Bolos forward, there were a number of practitioners who continued in the same vein in the examination of chemical processes. Strathern describes one such effort, made by a certain Zosimos of Panopolis around 300 A.D., as follows.

> Zosimos compiled an encyclopedia of alchemy in twenty-eight volumes, one for each letter of the Greek alphabet . . . The contents of this encyclopedia bear a similarly loose attachment to reality, being written in a style by turns cryptic, mystical, and nebulous. Symbolic formulae and coded experimental instructions abound. Just occasionally the clouds of mysticism and metaphor part to reveal a glimpse of something less vaporous. At one point Zosimos defines alchemy as the study of 'the composition of waters, movement, growth, embodying and disembodying, drawing the spirits from bodies and bonding the spirits within bodies.'

Because Strathern does not follow Kuhnian paradigm model, he views the work of someone like Zosimos from the standpoint of what we might call "objective error", meaning that he believes the modern position is objectively true, and would have been available to any observer or participant along the way, if they had only been smart enough to see it and not deluded. The Kuhnian position rejects this approach. Instead, the Kuhnian paradigm views Zosimos as being in "relative error", understanding that we judge his work by the stable paradigms that we rely on today. In other words, we would still view Zosimos as being in error, but for different (yet important) reasons. From the Kuhnian perspective we take account of the fact that our point of view is always conditioned by the existence of our paradigms, and have concluded, based on the record of thought to date, that no human has direct access to an overarching objective paradigm. We apply this limitation to ourselves, just as much as to any other participant in the line of history.

Of course, we today have no reason to doubt that Mendeleyev's periodic table of elements is correct and every reason to believe it is true; thus, it provides a stable vantage point. When we look back at the activities of prior alchemists and chemists, we must continue to view their work through the lens Mendeleyev established, unless and until we have good reason to doubt his work and see its undisclosed error. The difference between the Kuhnian perspective and the perspective of someone like Strathern is that Strathern seems to believe Mendeleyev's Periodic Table of Elements is objectively true for all time, and always could have been discovered, whereas the Kuhnian says that it could have only ripened into being after other conditions had been in place for some time. The Kuhnian sees a dynamic unfolding of paradigms over time; the conventionalist sees an objective, "eternal" paradigm just waiting to be clearly perceived by those who are smart enough to do so—the Kuhnian rejects this as a false conceit.

Thus, from the Kuhnian perspective, no matter how smart someone was, they could have never seen the Periodic Table prior to the latter half of the 19th Century, because an array of forces that were greater than the power of any individual observer had to first come into alignment. A conventionalist like Strathern would say that the people who didn't see Periodic Table before were simply ignorant, whereas we today are smart; someone who follows the Kuhnian approach, however, would never think in such a way because she understands that more factors than just the brilliance of a single observer are at play. In any event, we can say that these pre-scientific chemists were familiar with basic processes such as distillation, filtration and solution, even if it looks to us like they had no idea what they were really doing. They may have had some strange, mystical explanations for their work, but those reasons look wrong to us, even though the processes they were using seemed to actually work.

Another benchmark in the alchemical development of chemistry was the work of 8th Century Arab intellectual, Jabir al-Hayyan. Jabir worked under the Aristotelian doctrine of four basic elements, but modified it and also theorized the existence of chemical catalysts. Jabir was also an experimentalist. Strathern says that Jabir "managed to prepare weak solutions of nitric acid . . . Here were being assembled the simple ingredients for any basic chemistry kit." Jabir codified his knowledge in a text called *The Sum of Perfection*, which Strathern says included "a survey of all his vast alchemical and chemical knowledge." This line of work was followed by another Arabic chemist, Al-Razi, who published a book entitled, *The Secret of Secrets*, which continued in this same vein. Strathern says that one of the most interesting aspects of *Secrets* "is devoted to substances, and includes a long list of known chemicals and minerals. Here Al-Razi was the first to classify substances as either animal, vegetable, or mineral. He also produces a list of the different types of materials used by alchemists." In the 10th Century, the work of the Persian philosopher Avicenna moved this type of inquiry into the realm of medical science. Avicenna believed that "mineral or chemical remedies were far superior to the herbs and old wives' tales that had been prevalent since time immemorial," and applied chemical knowledge and analysis to develop a pharmacopoeia.

The works of the Arab thinkers eventually made their way to Medieval Europe. The first major European figure in the field of alchemy was Albert Magnus, who was born around 1200 in southern Germany. Drawing on the works of the Arabs and Greeks, this pre-scientific field shifted to an abstract search for truth, which eventually turned into a search to uncover the "Philosopher's Stone", as well as the search for a method to transmute other substances into gold. The Philosopher's Stone was described by the fourteenth-century alchemist Arnold of Villanova as follows: "There exists in Nature a certain pure substance, which when discovered and brought by Art to its perfect state, will convert to perfection all imperfect bodies that it touches."

Although the Philosopher's Stone obviously sounds to us today like something even less believable than the ether, this idea was an important motivating force in the drive to break down and study chemical processes. Strathern quotes the 19th Century German chemist Justus Lieberg on this point as follows. "The finest imagination in the world could not have conceived of a better idea than the 'philosopher's' stone to inspire the minds and faculties of men. Without it, chemistry would not be what it is today. In order to discover that no such thing as the philosophers' stone existed, it was necessary to ransack and analyse every substance known on earth. And in precisely this lay its miraculous influence." Strathern outlines the particular way in which this search for this imaginary inferred object led to real paradigms which later found a new place within the Periodic Table.

> . . . these alchemists were also laying the theoretical foundations of chemistry as we know it. In their ravening search for the philosophers' stone the 14th century alchemists became the first to understand the nature of acids. The only acid known to the ancients had been the weak acetic acid of vinegar. In the 8th century Jabir had prepared a weak solution of nitric acid and other Arab alchemists discovered that distilling vinegar made a stronger acetic acid. But even strong

> acetic acid was hardly corrosive. It didn't appear to be possessed of much reactive power. Then came the breakthrough. Just after 1300 [an alchemist known as] the False Geber discovered vitriol better known to us as sulphuric acid. Here was a liquid which appeared to dissolve erode our react with almost everything! This has been called the most significant chemical advance since the discovery of how to produce iron from its ore around 3000 years previously.

Under the Kuhnian paradigm outlook, this kind of thing is not surprising. The paradigm behind the search for the philosopher's stone might be completely fantastical, but because this drive focuses and limits the experimental attention to a narrower band of phenomena, it ultimately should produce some kind of knowledge. Under the Kuhnian analysis, the ideas behind the paradigms can be "wrong" while the paradigm itself still works and does things that are real. Although an alchemical paradigm would never explain how sulphuric acid is actually forming from the perspective we hold in 2025, it can still serve as a vehicle through which real scientific products are manufactured and explored.

Other discoveries continued in this same vein: strong nitric acid, which they called "strong water", the "water of life", a.k.a. alcohol, and so on and so forth. The drive to find the secret to transmuting other substances into gold, or the philosophers' stone, gave humans an impetus to pursue certain lines of study. Strathern describes the general attitude created by these fields of inquiry as follows,

> Besides being a rational soul, man was also a creator. But in order to exercise this divine power he first had to discover the secrets of nature this could be done only by torturing nature—by subjecting it to fire, dissolving it with strong waters, and subjecting it to other alchemical processes such as distillation. Success in these manipulations turned man into a God possessed of eternal life. It gave him power over matter, untold riches, and the ability to turn the world into a paradise. It is not difficult to recognize in this the vision which was to evolve into the central belief of science. Through technological innovation, experiment, and scientific thinking, man could rule nature by transforming it and imposing his will upon the world. According to the historian of science L Pierce Williams, "This is essentially the modern view of science and it should be emphasized that

it occurs only in Western civilization. It is probably this attitude that permitted the West to surpass the East, after centuries of inferiority in the exploitation of the physical world." In the midst of the most recalcitrant alchemy lay the seeds of the scientific fantasy which continues to propel the world today (whether we like it or not).

As noted above, there was a substantive shift in the practice of European alchemy with the work of Paracelsus, which then transformed in a pivotal way with the work of Boyle. But would the type of chemistry practiced by Mendeleyev even been possible without the long history of error that we see with the practice of alchemy? Perhaps we can say that alchemy was a necessary, but not sufficient condition for modern chemistry because it provided a level of real motivation that would be difficult to otherwise summon during that time. Maybe it was necessary to be powered by erroneous idea of such a nature that was appropriate to that era of human development. Perhaps it is unlikely that people would have even had the motivation necessary to investigate nature at that detail if they were just searching for things like chlorine, hydrogen, mercury, or other elements as we know them today. Would the investigations which were powered by the irrational desire to transmute gold or to find the philosopher stone even have arisen in any other way?

Of course, all of this is nothing more than rational speculation which we have no way of proving either way, since the passage of time has now laid a fixed course. The point of this exercise in thought is simply to see ways in which alien paradigms relate to each other in the real world. The paradigm of alchemy is alien to that of modern chemistry, but there are important points at which they are connected. The actual dynamics of this evolving process are not as simple as people like Strathern would have them. It is not simply the case that they were dumb or wrong and we are smart or right. We must, of course, rely upon the stable paradigms that are given to us, and those paradigms are stable when they appear to work in the real world. But unfortunately, many mistake this relative stability for enduring objectivity. Because of this viewpoint, they fail to see the dynamic and subtle ways in which alien paradigms relate to one another, making it extremely difficult to entertain the kind of thinking that will be necessary to envision the next level of future alien paradigms.

Consider the ways in which the paradigms of European alchemy and Mendeleyev are alien to one another. First, we can see that prior to Boyle, there was a way in which European alchemy deployed an approach that looked for more basic elements in nature which could be broken down or reformulated to transmute gold or find the philosopher's stone. In this respect, the paradigms are similar. But Boyle's approach of detailing chemical process in a scientifically written manner in order to simply understand them and socialize them, is radically different than the mystical and secretive approach pursued by the alchemists. This slight shift in approach is alien to that of alchemy and opened up an entirely new approach which allowed for the independent validation of results, creating an entirely new observational regime which developed over time.

This independent observational regime which Boyle's work allowed was core to the validation of Mendeleyev's paradigm as it provided the only vehicle through which the predictive power of the Periodic Table could be tested. Mendeleyev himself couldn't be the one to resolve the anomalies in the initial version of the Periodic Table on his own in order for the mechanics of the model to be validated. Only someone independently resolving those anomalies would be sufficient to convince others that the paradigm was real, and not simply a figment of Mendeleyev's mad Russian imagination. This independent process was necessary to instill the scientific conviction necessary for others to keep looking at the new paradigm and assemble the human team to start working on it. Without that conviction, the whole thing could simply be written off and thrown in the trash as nothing more than the ravings of a Russian lunatic.

We see that the Periodic Table was not simply a description and ordering of the chemical elements, as it possessed a certain predictive power regarding the elements which was completely new and unprecedented. This kind of approach, whereby the application of a paradigm of knowledge regarding chemical elements would lead to very precise information regarding the details of unknown elements, was completely and totally alien to that of European alchemy. A European alchemist could never dream of such a thing (nor, for that matter, would they even probably want to, as such a "modern" scientific goal would not likely be the type of thing that would naturally come to their minds or even interest them in the same way it interests us).

We can look at this issue in another way. For us, Mendeleyev's periodic table of elements appears to be a very comprehensive paradigm. It not only explains a wide range of phenomenon, but it also predicts a wide range of phenomenon that appear to us to be occurring in the real world. If there are serious and pervasive errors in the periodic table of elements, they are not apparent to us at this point in time. However, looking back upon the historical record, we can see that other humans in previous areas felt exactly the same way about their paradigms as well. In many of these cases, such as the field of chemistry and the Copernican revolution, we today can see the error which they could not. But it wasn't necessarily because they were "dumber" than us that they couldn't see these things. The truth is far more subtle than that and concerns the ways in which knowledge itself unfolds over time in relation to the limited lifetime of a human observer.

The record of human thought implies very strongly that humans 500 years from now would see a whole range of errors in the way we view the world that would be extremely difficult if not impossible for us to see. This is where the concept of alien paradigms comes into play. People that accept the existence of a whole range of alien paradigms will not be surprised by this and will accept that the long-range contours of alien paradigms will be inaccessible. This leads to an operating assumption that there will be a whole range of new and unknown paradigms in existence in the future which no human today could even imagine, and inclines us to adopt a certain attitude of openness to alien possibilities today so that we can be the ones who start to pull on the threads of these new paradigms and set the process in motion. We understand that we may not be around long enough to see the whole picture emerge and we understand our role in this process as one small piece of a much larger historical enterprise—an enterprise whose legacy far exceeds the significance of any particular observer. It is this open attitude that distinguishes the leaders and pioneers from the dying dinosaurs.

Beyond that, acceptance of the alien paradigm theory inclines us to also believe that even if a human today did correctly intuit or imagine such an alien paradigm, it would take some real time to help give it real meaning, and the lack of immediate meaning would not be a signal that the paradigm is necessarily wrong. Again, think back to the example of the Copernican paradigm. Copernicus' paradigm was clearly alien in relation of the world of conventional thought in this time. The paradigm made perfect and complete sense on the narrow issue of how it was that the earth moved in space, thus providing an explanatory framework for the motions of celestial bodies. However, Copernicus' paradigm failed to explain many other anomalies it created, which was a large factor as to why it was not readily accepted and was in fact resisted. But it would be unfair and irrational for us to expect that Copernicus would have been required to also figure out Newtonian physics before he could work on his paradigm. Moreover, if Copernicus had listened to his detractors and refused to accept the findings of his paradigm due to its lack of comprehensive explanatory power, then no progress would have been made and he would have just thrown it in the trash because it wasn't complete enough for his critics. Ironically, this refusal to pursue the matter because something like Newtonian physics had not yet been developed, would have then prevented the development of Newtonian physics at a later date. Thank God Copernicus didn't do that!

The same type of dilemma, on a more limited scale, confronted Mendeleyev because his work contained apparent anomalies as well. In both of these cases we can see, however, that it was necessary to ignore the overall incompleteness of the paradigm in order to move things forward. This is a sign that an alien paradigm may be in the process of forming. We can also see that if no effort had been made to move in a positive direction, then the later paradigm which then helped fill in the gaps and explain things which were mysterious to both Copernicus and Mendeleyev would've never been discovered.

Paradoxes abound here, as this work is not for the comfort-craving conventionalist. There are at least two possible reactions to them. One way is to simply categorize their mysterious aspects as symptoms of delusion. In this approach, we would say that there is nothing mysterious about the world, but that the world only appears mysterious to us when we are ignorant. When our ignorance is dispelled through the application of reason, science, and logic, then things which appeared mysterious in the past no longer appear so. Under this paradigm of explanation, a person believes that there is an objective paradigm of truth, which humans are able to apprehend and perceive and that prior eras were just being stupid when they couldn't see it like we do. This is very similar to assumptions underlying the universal scientific community hypothesis, which was described above. This is the conventional viewpoint of the human relation to truth.

The alternative approach to this is the one described by Kuhn. Under the Kuhnian approach, we realize that our knowledge is constructed of nothing more than a series of relative paradigms that last for some period of time. These paradigms emerge, dissolve, change, transform, and eventually disappear. For reasons that are not entirely clear to us, some paradigms appear to work for limited periods of time and then either break down or change the longer they are used. Some paradigms appear to be much more long-lasting than others. For example, the overall Copernican paradigm still holds valid today and from our perspective, we see no good reason why it would not to continue to endure forever. This does not mean that it will hold true indefinitely, as we cannot predict how humans in 10,000 years, 100,000 years, or 1 million years will see the world. Similarly, applying Nimoy's Dictum, we cannot say that humans existing in this time range wouldn't see a number of undisclosed errors in Mendeleyev's Periodic Table, which we today simply cannot see, just as the European alchemists could not have seen our Periodic Table. We may not even be able to imagine what such errors would look like. And even if we could accurately imagine such errors, they would only appear to us to be visionary at this point in time, as we would not have a way to physically explore, examine, and detect them. They would remain completely alien to us, even if they were just as real as the things we see today.

Under the analysis of relative paradigm perspectives, we come to accept that proving something as a "real experience" to a human being is only one level of the analysis, but is not the end all/be all of anything because we see how prior eras felt their delusions to be perfectly real as well.

Of course, the experience of the real does give us certain forms of information. For example, if one believes that there is such a thing as an objective paradigm of truth that humans actively interact with, and believe that things like the Periodic Table of elements are objectively real in all times and all places in precisely the same ways, then such a person would be more likely to believe that any type of alien consciousness would see the Periodic Table, and the aspect of reality it represents, in the same way. This would place an *a priori* limiting constraint on an alien paradigm. By contrast if one understands that the Periodic Table of elements certainly *seems* correct but can only be *relatively true,* one possesses an openness of mind that acknowledges there could be unimaginable ways in which the evidence and information which is presented in the Periodic Table development could be seen. This does not incline us to denigrate or reject our own experience but merely provides it a context which is more in alignment with the long-range trends in the record of human thought itself. This makes it more likely that we will be able to actually adapt to alien paradigms and fill in their details much more quickly. Instead of acting like Lord Kelvin or the adherents of SETI, we will be more like Roentgen or Mendeleyev and be ready to flow with the future.

5.3 *Alien anomalies as pre-scientific phenomena*

In 2025, just as was the case in 1947, the study of alien anomalies is in a "pre-scientific" phase of development. This means that no one who takes these issues seriously would be demanding extraordinary proof to materialize as if reality was some kind of science fiction movie, but rather to follow a patient and steady approach. Just as the medical clinicians of the 18th century abandoned a fixation on the theory of nosology and focused on what was actually in front of them, we today will be more interested in collecting observational data, taking measurements, and learning about the anomalies themselves instead of theorizing about things like Dyson spheres, alien megastructures, and deep space radio signals. Of course, this would require us to collect such information on a large scale, in real time—something which would take a new kind of scientific effort and effort which is very distinguishable from the continued practice of spending an enormous amount of time on analyzing past occurrence (for which no preparations at detection had been made in advance) instead of placing collective efforts on being prepared in advance to observe future anomalies such as the tic tac.

Whether the 1947 Air Force knew it or not, it appears to have instinctively hit on the type of approach that is consistent with an approach to alien anomalies as pre-scientific phenomena. The scientists of the 1940's, being under no pressure to understand or deal with alien anomalies, could avoid them altogether or engage in theoretical speculation about what they could be; the Air Force, by contrast, was forced to admit them into study and intuitively took the approach that is consistent with Kuhnian paradigm theory. Of course, the military abandoned the Project Sign approach in favor of suppression, but had the initial path been followed, in 2025 we would have been the inheritors of over 60 years of active, open inquiry, as opposed to suppression, denial, and evasion.

Treating alien anomalies as pre-scientific phenomena requires certain initial measures to see them clearly in the field of thought. First, we must have an accurate understanding of paradigm theory and development, so that we can properly weigh the value of the evidence they present and see the real role that such evidence will play in normal scientific process. This, among other things, means that we will abandon Sagan's Dictum. Second, we will have a goal of working with alien anomalies as scientific objects, instead of religious, spiritual, or fantasy ones. Third, we will agree that the "invisible" nature of alien anomalies will not be resolved by waiting around for the evidence to drop out of the sky like we are in a science fiction fantasy movie, but will be resolved in the same type of way that we resolved the invisibility of other objects (such as the atom or the ether).

5.3.1 *Identifying the neurotic side effects of alien anomalies*

Since there has been 60+ years of hardened resistance to, and distortion of, alien anomalies under the safe distance paradigm, and it has become so embedded and entrenched within conventional scientific thought, the path towards genuine scientific treatment of these anomalies makes it necessary to directly address the neurotic and inhibitory effects the safe distance paradigm has had on the field of collective thought so that people can be aware of these effects and not succumb to them.

For starters, it is difficult to fathom why someone who is genuinely interested in learning about alien intelligence would so strongly reject the proposition that a near earth study of their potential presence would be just as valid as a search for deep space radio signals. After all, why not support as many avenues as possible? But adherents to the SETI paradigm are not just adverse to a near-earth study, they are hostile and dismissive to its mere possibility as if it presents some kind of grave threat to their citadels of security and stasis. This reaction bears the hallmarks of a neurotic reaction. "In neurosis," writes David Shapiro in *Psychotherapy of Neurotic Character* (1989), "the personality reacts against itself. It seems that the person has tendencies that his own character cannot tolerate and reacts against, with remarkable consequences . . . such a reaction by the personality against itself leaves the person who experiences it estranged, cut off, from himself in certain ways." In a neurotic situation, the subject "does not know what he wants or what he wants to do. He does not know what he feels; or sometimes he knows he has strong feelings or reactions, but they seem strange to him, as in the case of phobias, mysteriously at odds with his judgment, attitude or common sense." In order to resolve a neurotic conflict within the individual, they must be restored to an "awareness of their own wishes and feelings, and their awareness, as well, of their reactions against those wishes and feelings. In this way . . . [they can] transform an intrinsically unresolvable neurotic conflict, a reflexive and inaccessible reaction against the self, into an ordinary conflict of possibilities."

The fixation on deep space radio signals, alien megastructures, Dyson spheres, and other ideas of a similar nature, appear as signs of a neurotic reaction in this context because none of them can become ordinary objects of scientific action and discourse. On the one hand, this expresses a genuine desire to learn about aliens; on the other hand, it renders the desire impossible to satisfy. This is not a real scientific approach, but a pseudo-scientific fantasy, one that is fundamentally untestable in any realistic way. This is likely a byproduct of what Jung identified as the "numinous effect" of the topic of alien life on the human subconscious mind in *Flying Saucers*.

> At various times all sorts of other projections have appeared in the heavens besides the saucers. This particular projection together with its psychological context, the rumor, is specific to our age and highly characteristic of it. The dominating idea of a mediator and god who became man, after having thrust the old polytheistic beliefs into the background, is now in its turn on the point of evaporating . . . No Christian will contest the importance of a belief like that of the mediator, nor will he deny the consequences which the loss of it entails. So powerful an idea reflects a profound psychic need which does not simply disappear when the expression of its ceases to be valid. What happens to the energy that once kept the idea alive and dominant over the psyche? A political, social, philosophical, and religious conflict of unprecedented proportions has split the consciousness of our age. When such tremendous opposites split asunder, we may expect with

certainty that the need for a savior will make itself felt. Experience has amply confirmed that, in the psyche as in nature, a tension of opposites creates a potential which may express itself at any time in a manifestation of energy. Between above and below flows the waterfall, and between hot and cold there is a turbulent exchange of molecules. Similarly, between the psychic opposites there is generated a "uniting symbol," at first unconscious. This process is running its course in the unconscious of modern man. Between opposites there arises spontaneously a symbol of unity and wholeness, no matter whether it reaches consciousness or not. Should something extraordinary or impressive then occur in the outside world, be it human personality, a thing, or an idea, the unconscious content can project itself upon it, thereby investing the projection carrier with numinous and mythical powers. Thanks to its numinosity, the projection carrier has a highly suggestive effect and grows into a savior myth whose basic features have repeated countless times . . . In recent times, however, and just at the moment when the eyes of mankind are turned towards the heavens, partly on account of their fantasies about possible space-ships, and partly in a figurative sense because their earthly existence is threatened, unconscious contents have projected themselves on these inexplicable heavenly phenomena and given them a significance they in no way deserve. Since they seem to have appeared more frequently after the second World War than before, it may be that they are synchronistic phenomena or "meaningful coincidences". The psychic situation of mankind and the Ufo phenomenon as a physical reality bear no recognizable causal relationship to one another, but they seem to coincide in a meaningful manner.[61] The meaningful connection is the product on the one hand of projection and on the other of round and cylindrical forms which embody the projected meaning and have always symbolized the union of opposites.

(emphasis supplied)

---

[61] This is another way in which to recognize that real aliens have a real distribution in time and space that has absolutely nothing to do with our own processes and ideas about them since aliens have developed under independent timescales and conditions of which we know nothing.

Jung noted in *Flying Saucers* that the UFO topic has a powerful numinous effect on the human mind. Other scientific topics, such as quantum entanglement or relativity, might have strong effects on individual communities of scientists given their strong commitments to certain paradigms (Einstein, for instance, was very resistant to quantum entanglement), but aliens seem to have a powerful effect on the populace at large and many scientists in particular. As Jung noted, many people instinctively want the reports about UFOs to be true, and the reports have always seemed to provoke fantasies. Why is this the case? Why do we find the possibility of alien to human contact so intoxicating, the cause of so much emotional disturbance or excitement?

As we have seen since Jung's work in 1958, the overwhelming popularity of alien culture in the science fiction market continues to grow, even though the scientific interest in the subject matter remains *de minimus.*[62] Jung saw in the strong provocation of such fantasies the expression of a deep yearning to be free from various forms of global distress and challenges facing human beings. "We have here a golden opportunity for seeing how a legend is formed, and how in a difficult and dark time for humanity a miraculous tale grows up of an attempted intervention by extra-terrestrial 'heavenly' powers—and this at the very time when human fantasy is seriously considering the possibility of space travel and of visiting or even invading other planets." Jung recognized that, regardless of whether UFOs are real or not, the reactions by humans to the very idea and possibility of them being real was, in and of itself, a real phenomenon from which we could gain valuable insights. Jung elucidates this position as follows,

> This attitude on the part of the overwhelming majority provides the most favourable basis for a projection, that is, for a manifestation of the unconscious background. Undeterred by rationalistic criticism, it thrusts itself to the foreground in the form of symbolic rumor, accompanied and reinforced by the appropriate visions, and thus activates an archetype that has always expressed order, deliverance, salvation, and wholeness. It is characteristic of our time that the archetype,

---

[62] Just the market for science fiction books alone is supposed to surpass 12 billion dollars in the next several years, and the market for movies and video games involving aliens is even greater. At what point in time will we spend a similar amount of money on a scientific study of these alien anomalies, instead of simply regarding them as pure entertainment?

> in contrast to its previous manifestations, should now take the form of an object, a technological construction, in order to avoid the odiousness of mythological personification. Anything that looks technological goes down without difficulty with modern man. The possibility of space travel has made the unpopular idea of a metaphysical intervention much more acceptable. The apparent weightlessness of UFOs is, of course, rather hard to digest, but then our own physicists have discovered so many things that border on the miraculous: why should not more advanced star dwellers have discovered a way to counteract gravitation and reach the speed of light, if not more?
>
> . . . I still remember, with misgivings, the time when I was convinced that something heavier than air could not fly, only to be taught a painful lesson. The apparently physical nature of the Ufos creates such insoluble puzzles for even the best brains, and on the other hand has built up such an impressive legend, that one feels tempted to take them as ninety-nine percent psychic product and subject them accordingly to the usual psychological interpretation . . . For primitive man any object, for instance an old tin that has been thrown away, can suddenly assume the importance of a fetish. The effect is obviously not inherent in the tin, but is a psychic product.

It is necessary to acknowledge the neurotic side effects these alien anomalies have on human beings if we want to avoid succumbing to such neurosis. Neurotic effects cause distortions when they are not consciously acknowledged by the subject but are dispelled when made an object of conscious examination and insight. While one cannot describe exactly how a neurotic effect is dispelled through this conscious process (as it is different based on the individual's temperament, character, and general abilities), one can say that a refusal to at least acknowledge the neurotic side effect will certainly make it worse. By acknowledging these effects and working through them in a rational way, however, the individual may have a chance at seeing the anomalies in a normal kind of way.

As Dr. Shaprio stated in *Psychotherapy of Neurotic Character*, we know we have resolved a neurotic conflict when whatever is bothering us has been turned into an "ordinary conflict of possibilities". In the context under discussion here, this means that we stop wasting time on unproductive speculative activities, or on making alien anomalies more spooky, mysterious, or ominous, and instead simply treating the anomalies as pre-scientific phenomena within the larger context of scientific process itself, and that we start spending as much money on scientific examinations of alien anomalies as we spend on science fiction fantasy entertainment about them. Once alien anomalies are treated as pre-scientific phenomena, they are being handled in an ordinary way from the standpoint of scientific process itself.

Think of it this way. Imagine that when Von Frisch was studying his bees he was tripping on LSD the whole time and obsessed with the idea that the bees were carrying special messages which could save humanity, instead of just studying them with a basic air of curiosity to see what he could learn. We would not only think him mad, but he would also likely not have developed any real knowledge. So why then do we continue to take this type of attitude towards alien anomalies today? When will we get tired of learning nothing new or real about them and just tripping out on them all the time as mystical objects?

5.3.2 *Fear of the uncanny*

There can be no doubt that the subject of aliens arouses great interest and strong emotional reactions among humans, regardless of whether aliens or real or whether humans are merely projecting subconscious fantasies into thin air. And although many people claim that they would be dying to meet or contact aliens in real life, there is certainly a possibility that the experience of genuine contact might also carry with it a subconscious element of supernatural horror; thus, pulling the human mind towards something which it also simultaneously fears and setting the stage for a genuine neurotic crisis. In *The Conspiracy Against the Human Race* (2010), writer and philosopher Thomas Ligotti identifies elements of supernatural horror to include "the paradox of the flesh . . . something which should not be, and yet is," as well as elements of the appearance of "the uncanny", disruptions of normality, suggestions of "unknown modes of being", "cosmic disillusionment", or the confrontation with "things that should not be". Clearly, the actual presence of aliens in the human world would check all of these boxes, especially when aliens are approached as potentially real objects of scientific inquiry.[63]

> In the literature of supernatural horror, a familiar storyline is that of a character who encounters a paradox *in the flesh*, so to speak, and must face down or collapse in horror before this ontological perversion—something which should not be, and yet is. Most fabled specimens of a living paradox are the "undead", those walking cadavers greedy for an eternal presence on earth. But whether their existence should go on unendingly or be cut short by a stake in the heart is not germane to the matter at hand. What is exceedingly material resides in the supernatural horror that such beings could exist in their impossible way for an instant. Other examples of paradox and supernatural horror congealing together are inanimate things guilty of infractions against their nature . . . a puppet possessed of life would exemplify such a horror, because it would negate all conceptions of a natural physicalism and affirm a metaphysics of chaos and nightmare.

We could certainly add "embodied alien presence" to this catalogue of horrors. In another place in his work Ligotti refers to the sense that "behind the scenes of life are baleful agencies that govern the world of supernatural horror" which could be an apt description of the anxiety that is expressed in Fermi's paradox: aliens should be here, so why am I not seeing them? "Such is the motif of supernatural horror," Ligotti tells us, "Something terrible in its being comes forward and makes its claim as a shareholder in our reality, or what we think is our reality and ours alone." This kind of confrontation unsettles the default idea that the earth is our and ours alone, a narrative structure that is repeated through all of the major religious of planet earth and has been ingrained in us all since the earliest days of our prehistory as sentient beings.

In another part of *Conspiracy*, Ligotti cites a passage from an essay entitled "The Eyes that Never Blink", which restates this possibility of existential fear.

---

[63] Professor Madden discusses the element of the "uncanny" throughout his work in *Unidentified Flying Hyperobject*.

> Mist on a lake, fog in thick woods, a golden light shining on wet stones—such sights make it all very easy. Something lives in the lake, rustles through the woods, inhabits the stones or the earth beneath them. Whatever it may be, this *something* lies just out of sight, but not out of vision for the eyes that never blink. In the right surroundings our entire being is made of eyes that dilate to witness the haunting of the universe . . . at any time and in any place, our bunkers of banality begin to rumble. You see, even in a stronghold of our fellow beings we may be subject to abnormal fears that would land us in an asylum if we voiced them to another. Did we just feel some presence that does not belong among us? Do our eyes see something in the corner of that room in which we wait for we know not what?

When we begin to break down the safe distance that separates us from alien anomalies, stepping outside the safe confines of science fiction, and into the realm of the actual, it would be natural to expect that some neurotic aversion would arise. This would be especially true for scientists who take their work seriously, and for whom scientific activity is a genuine passion in which they place great faith. For these people, to study aliens as real objects of scientific inquiry is a serious task, and one which may arouse strong feelings of fear and cosmic dread. Indeed, it could even be seen as an act of scientific bravery to begin to think about aliens as real objects of science for such people. Accordingly, these emotional reactions should be taken seriously and not lightly dismissed, for they speak to exactly the type of serious commitment that is necessary if genuine progress is to be made. Again, working through these subconscious reactions in a rational and conscious way may allow an individual to see alien anomalies more clearly and normally for what they really are.

5.3.3 *The "home effect"*

We cannot underestimate the subconscious effect that an alien presence would have on many people. As we discussed in section 2.1, conventional categories of thought allow us to place phenomenon in a settled order, permitting us to either move on to something else or to work with the phenomenon in a relatable way. When items fail to have a place in a settled order, they are a cause of discomfort and concern, and there's a natural tendency on our part to want to fit them into a settled order. It is almost as if something is out of place or doesn't feel right, which we want to correct. Accepting the reality of an actual alien presence in one's world is uncomfortable precisely because it is not a settled conventional category at present (i.e., an accepted "ultimate cause"). Anybody who has ever been startled by an unwelcome presence understands the unsettling nature that that type of experience creates. We can think of many different common examples of when an unwelcomed or unforeseen presence intrudes on someone's life and causes them discomfort. For example, one can think of having one's home invaded by a burglar. One can think of a situation where one is enjoying a peaceful day at a beach and a shark suddenly appears in one's range. One could think of hiking in the woods and suddenly coming upon a bear. We can think of a lot of different examples from common experience where the presence of an unforeseen presence creates a startling reaction. The effect that these direct experiences have on people is discussed in greater detail in the next section of this work, *Alien Experiences*.

But here we are contemplating a situation where the individual is not necessarily having a direct experience of an alien embodied presence but is engaging with the possibility of an alien presence in a real and thoughtful way. In other words, not simply contemplating it as a science fiction possibility, but actually contemplating it as a real possibility which could have a real effect on how they perceive the world; in other words, we are discussing the possibility of somebody changing their paradigm in how they look at alien presence, going from a state where they believe that an alien presence on earth is only a science fiction possibility, to one in which they accept it as a real possibility which has a startling effect on their life. This is a variation on the thought experiment that was suggested in **Section 2.9** ("the real, the unreal, and the 'Copernican Trap'"), with the example of the two people—one in scenario A where they are watching a movie about aliens and know it's a movie, and scenario B where the viewer of the movie has unknowingly been given LSD and doesn't realize that the movie is just a movie. The variation in this situation under consideration here is that we are simply contemplating this as an exercise in the field of thought; we are considering directly a reader of this work who may be an adherent of the SETI paradigm, but as they contemplate this issue and as they follow the arguments in this text are starting to question their previous assumption and now having a certain degree of openness to the possibility that alien presence on earth could be a real thing today. If a person genuinely reaches this level of thought within themselves and it is a genuine good faith reconsideration that might actually change their paradigms, their perceptions, and their future conduct, such a person may be experiencing a neurotic crisis of realizing that alien presence on earth is a real thing. This would naturally cause them to have an unsettling effect with respect to their subconscious feelings about the stability and settled nature of earth as their home planet. Many would prefer to avoid such uncomfortable thoughts and feelings altogether.

To such a person, we might have a number of questions. For example, instead of being overly concerned or overly worried about whether their idea of earth would be unsettled, we might simply wonder if that individual had some very strong subconscious beliefs about what makes earth feel home to them that were not examined until the moment when they seriously contemplated alien anomalies in this way. In other words, we might wonder if this person had a particular viewpoint of earth as belonging to humans only as the prime species and that belief gave them a certain degree of comfort and stability in their lives which is now being challenged. By contrast, it could be the case that individuals who are more open to the reality of alien anomalies intruding on earth don't feel such a strong need for comfort and reassurance in that way. They may have a different view of planet earth and a different view of why they feel comfortable and secure in planet earth that has nothing to do with whether it is ours and ours alone.

In any event, looking at these alien anomalies in a realistic way brings us to these questions. These types of questions are non-neurotic because they are asking us to inquire as to our genuine belief systems. We are not simply sitting back gazing up at the stars and wondering if are we alone, instead we are actively examining our own subconscious beliefs and ideas about earth, about how those ideas would be affected if we took the existence of alien presence as a real possibility, and understanding how our subconscious beliefs could cause our thinking to be distorted. Again, as mentioned repeatedly throughout this work, the conventional viewpoint today among science is that aliens do likely exist in the universe. The issue comes down to this: just as people in Copernicus's time understood that the new conventional truth that Copernicus expressed was probably right, the disconnect between that conventional truth and their lived experience made it difficult for them to fully embrace the new conventional truth that Copernicus was articulating and explore it in a productive way. It is getting over that internal feeling of absurdity, ambivalence, and neurosis, that leads to productive pathways of knowledge and brings new levels of interest, inquiry, excitement and exploration to our species. None of us today would praise or congratulate those who resisted the Copernican truth. Today we view it as unfortunate. Today we wish that more people in Copernicus's time had been open to the new conventional truth that he was articulating. So why shouldn't we take the same attitude today in relation to alien anomalies if reason and the circumstantial evidence are urging us to do so?

5.3.4 *Avoiding anti-Copernican errors*

As argued repeatedly in this work, it is quite possible that we today are subconsciously repeating errors with respect to alien anomalies that are similar in nature to what happened when people in Copernicus' time were confronted by the anomaly generated with his upsetting work. And even though one can understand and sympathize with this subconscious resistance, one must also note that the resistance appears to be fundamentally arbitrary in nature. And, indeed, while it would be extraordinary—or even terrifying—for us to find out that aliens were present on earth, that doesn't mean it would be scientifically "extraordinary". For all we know, there could be all kinds of intelligent alien beings zipping across the universe, doing things which we couldn't even imagine, or which wouldn't make any sense to us at all.[64] And as for why these aliens wouldn't make their presence visible and irrevocable to us, there could be many reasons. We could speculate about such matters, but at the end of the day we have no more reason to believe that they would announce themselves to us, than that they would conceal themselves from us and chose to silently haunt our planet.

When we are thinking about the possibility of actual alien beings, we must take into account the fact that we have been conditioned from time immemorial to believe a number of things that impact how we view aliens which we have no way of knowing are true or

---

[64] A fun example of a world which is completely alien to human beings is found in the animated series *Scavenger's Reign* (2023), which depicts humans interacting with a world that operates by an independent alien logic of life.

not. For example, most people would believe that aliens have never visited earth before—but how could we possibly know if that was true? The earth has been in existence for close to 4 billion years and gone through major catastrophes and upheavals during that time. While our knowledge of earth's history has increased dramatically in the past several hundred years, it is still spotty at best. We have no idea if aliens established any colonies, outposts, or settlements in the distant past. For all we know, aliens could have visited earth in prior eras, observed the activities here, and then left. Similarly, for all we know, aliens could have been observing earth, "patiently waiting" to see if intelligent life developed here, and then took a lower profile once it did so as not to disturb it. There are so many possibilities here, and we have no way of knowing which are more plausible than any other. The best we can say is that we have no direct and substantial evidence of prior alien activities on planet earth, but applying Nimoy's Dictum again leads to the inescapable conclusion that we cannot conclusively rule such possibilities out *in toto*.[65]

Assume for the sake of argument that in the past there were alien civilizations or alien travelers who saw earth as they were going through the universe on travels which had nothing to do with us. Let us say such an occurrence happened 150 million years ago, during the timeframe of Concestor 13 as referred to in **Section 3.3.1**. As noted in that section of the work, conventional science informs us that approximately 90 to 100 million years ago our predecessor being was a nocturnal non-scientific shrew-like creature that is referred to as Concestor 13, and that we later evolved from that evolutionary platform. So let us assume that a hundred million years ago there was a group of aliens traveling through the universe and they saw planet earth through some detection system that they have and decided to pay it a visit. This scenario of course sounds like something out of a science fiction movie to us because it is so counter to our actual lived experience, but from an objective scientific standpoint there is nothing scientifically extraordinary about this one way or the other. It may or may not be true, but we are not in a position to determine if it is

---

[65] There is a school of thought referred to as the "ancient aliens" or "alien astronaut" hypotheses which would point to certain anomalous evidence in the archaeological or historical record as indicating prior alien presence on earth. While nothing said here should be construed to discount such hypotheses, none of them has reached the level of mainstream scientific acceptance to date.

extraordinary or not because we have no information one way or the other on the actual distribution of aliens in the universe or their actual technological capabilities. However, the application of basic rationality in the context of Nimoy's Dictum leads us to the conclusion that we cannot rule such a thing out conclusively and it remains an open possibility within the field of cosmological time.[66]

So that may be an exercise in conventional thinking, but what does our human experience show us? Our human experience does not show us this at all. There are some people that that believe there is evidence in the human record of alien presence at prior times on earth, but without delving into the minutiae and the arguments that those folks present, what we can say is that those suppositions simply are not part of the conventional corpus of knowledge. Perhaps they are right and perhaps conventional knowledge is wrong, but they are simply not a part of conventional knowledge. This means that they are therefore not a part of conventional human experience.

Again, return back to **Section 1.3** ("the Copernican revision") and review the statements in the poems that are cited, as well as the statements of Jean Bodin and others, that provide a narrative readout of the conventional knowledge they had at their time. Looking back at the analysis presented in this work, we now see how those are paradigms of knowledge that colored and informed their experience at the time—but they were not the objective

---

[66] Conventional science posits the existence of the universe to be approximately 15 billion years old. It also posits that planet earth is approximately 4 billion years old. Conventional science tells us that 4 billion years ago humans in their present form did not exist on planet earth, which means that within 4 billion years a planet formed and produced the scientific beings that we know today as humans. Within this conventional understanding, the universe would have ample time for intelligent beings to arise on other planets in the universe. Also as noted in **Section 3.3.1**, the human record of our own technological development shows us that an intelligent species such as humans can go from extremely crude forms of technology to our current level of sophistication in a matter of 6,000 years. These facts show us that there is more than ample time within the cosmological order itself for this imagined scenario of an alien being or set of beings visiting earth 100 million years ago to have actually happened. Again, this is not science fiction speculation as we are simply applying conventional reasoning and seeing the strange and startling possibilities that it reveals.

measures of truth that they thought they were. Similarly, we can visualize a reader of this work who is an adherent of the SETI paradigm and finds a suggestion that aliens could have visited earth in the prior time to be absurd and fantastical. They would naturally point to the lack of physical evidence in the record as affirming their belief, but we would ask that person why would there be physical evidence in the record at all. For if alien civilizations had visited here, they could have had very little physical impact on the earth, or left no traces at all. They could have built some cities or structures and then left or they might not have had a need for such structures, especially if they had a very different relationship to time space and matter than we do, and especially if they had been if they had achieved a level of technological awareness that we are at today 10,000, 100,000 or a million years in the past (which, as we've noted previously, are all relatively reasonable time frames within cosmological space itself even though they are meaningless and imaginal to us because we measure out time at much smaller cycles). While we can work with larger timeframes this way in our thought, they are essentially imaginary or visionary from the standpoint of conventional experience. Thus, it is hard for us to relate to them in a personal way, in a way that is perhaps similar to the dilemma faced by those who were considering Copernicus' work in the 16th and 17th centuries.

At this point in the journey, conventional systems of knowledge will urge the observer to check out and go back to working in a more stable conventional field. However, those who are interested in exploring entirely new fields of thought and moving into entirely new realms of knowledge, will not ignore the struggle against feelings of absurdity, irrationality, and the tension created between conventional knowledge, conventional experience and the new emerging alien paradigm. Of course, it is up to each individual to decide how they wish to move forward at this point in time. All we can say is this: the more who are willing to move forward, and the sooner they are willing to move forward together, the more likely it becomes that new systems of knowledge will be built, because systems of knowledge among humans are a form of a community activity which cannot be completely accomplished in isolation.

Another consideration which we must bear in mind as potentially distorting our view here is the relentless narcissism and solipsism of the human species in general. As E.O. Wilson stated in *The Meaning of Human Existence* (2014),

> We're addicted to anthropocentricity, bound to a bottomless fascination with ourselves and others of our kind. Even the best-educated live on an ad libitium diet of novels, movies, concerts, sports events, and gossip all designed to stir one or more of the relatively small range of emotions that diagnose *Homo sapiens*. Our stories about animals require human-like emotions and behavior understandable with well-worn guidebooks of human nature. We use endearing animal caricatures, including those of tigers and other ferocious predators, to teach children about other people.
>
> We are an insatiably curious species—provided the subjects are our personal selves and the people we know or would like to know. The behavior goes far back beyond our species in the evolution of the primate family tree. It has been observed, for example, that when caged monkeys are allowed to look outside at a variety of other objects, their first choice for attention is other monkeys.

Most groups of humans have developed mythologies placing themselves, and planet earth, at the center of a grand universal narrative. This relentless self-obsession easily translates into a belief that aliens would want to meet us, know us and be a part of our cosmically significant dramas. But is there any valid scientific basis for such claims? For all we know, the type of development and history that we see on planet earth could have played out many different times on many different planets throughout many different solar systems. In theory, there could be intelligent aliens who have seen something similar to the human drama on more than one occasion and may even find our obsession with our own universal self-importance to be boring. Or they might be interested in us in the same ways that we are interested in studying ants, bees, or birds. We simply do not know. Aliens that have seen other "primitive" life forms with consciousness such as ourselves may have different ethical or philosophical vies on how they should or should not interact with them. If we consider aliens as genuine objects within the field of thought (even though we have no scientific confirmation of their reality) we must take these various possibilities seriously. If we have distorted the measure of our own universal importance without being aware of it, it could cause us to carry a number of unconscious biases that may create distorted expectations about how aliens might react in relation to us.

While considerations of this nature might have no legitimate role with respect to objects of normal scientific inquiry, one could argue that arbitrary views about the mores, habits, and behavioral characteristics of aliens with respect to humans is a critical extra-scientific factor which must be properly accounted for and itself made a conscious object of study. Because the vast majority of scientists have assumed that aliens visiting earth would immediately land on the White House lawn and asked to be taken to our leader, they have taken the absence of such activity to mean conclusively that aliens aren't here. Most scientists cannot even conceive of the possibility that visitors from other planets might have a whole host of reasons for not injecting themselves directly into our affairs and may be content with casual or surreptitious observation of our activities, or even take more interest in other features of our planet and no interest in us at all. Just because we may think we are the main attraction on planet earth, doesn't mean an alien mind would see it that way. As Jung noted in one of the quotations cited above, "[t]he psychic situation of mankind and the Ufo phenomenon as a physical reality bear no recognizable causal relationship to one another, but they seem to coincide in a meaningful manner." For a species that assumes everything in the universe revolves around or relates to itself, this bias may take some time to work through. These types of "anti-Copernican" errors must be corrected in the field of thought first and foremost because they will distort all subsequent views and activities.

5.3.5 Correcting the barriers to paradigm formation

Generations of science fiction propaganda appear to have distorted our views of how intelligent scientific beings on other planets might actually evolve and how they might actually think and behave. While science fiction certainly plays a valuable and important role in the development of actual science, at the same time popular science fiction tends to heavily anthropomorphize the galactic order of life and projects human behaviors onto a wide variety of alien life. Indeed, if science fiction did not reflect back to us human dramas, we would find it to be of little interest and would probably never pay to see it. By contrast, a scientific curiosity is not motivated by these "lower" forms of interest, based on sensation or stimulation. As Bohm notes in *On Creativity*,

> Is it then that the scientist likes to solve puzzles? Does he want to get a "kick" out of meeting the challenge of explaining a natural process, by showing how it works? Of course, a scientist may often find this aspect of his work enjoyable. Nevertheless, such enjoyment has properly to come as a *by-product* of something else that goes much deeper than this. Indeed, if a scientist worked *mainly* in order to get hold of such pleasures and continue them as long as possible, his activity would be not only rather meaningless and trivial, but also contrary to what is needed for carrying out research effectively. For the recognition that one's ideas are false or on the wrong track (which is often crucial for making real progress) generally gives rise to extremely unpleasant feelings of disappointment and failure: to avoid these, the scientist whose *first* aim was pleasure would tend to overlook weak points in his work (as indeed does unfortunately tend to happen with surprising frequency).
>
> It seems, then, that the answer to the question of why the scientists are so deeply interested in their work is not to be found on such a superficial level. Scientists are seeking something that is much more significant to them than pleasure. One aspect of what this something might be can be indicated by noting that the search is ultimately aimed at the discovery of something *new* that had been previously *unknown*. But, of course, it is not merely the novel experience of working on something different and out of the ordinary that the scientist wants—this would indeed be little more than another kind of "kick". Rather, what he is really seeking is to learn something new that has a certain fundamental kind of significance: a hitherto unknown lawfulness in the order of nature, which exhibits *unity* in a *broad range* of phenomena. Thus, he wishes to find in the reality in which he lives a certain oneness and totality, or wholeness, constituting a kind of *harmony* that is felt to be beautiful.

If one finds this passage persuasive, one can see why an overexcited, numinous expectation regarding alien anomalies would be counterproductive to a sober, disinterested, scientific study of them. It may also steer us away from seeing the anomalies clearly in their own light. For example, if we become overexcited as to the possibilities that aliens would want to be our friends, we would overlook the possibility that alien life could be so alien to us that we might have a very difficult time even relating to it at first. Going back to the idea suggested in the image of Von Frisch tripping on LSD, imagine if Von Frisch was so desperate for connection that he perceived the development of an emotional bond between himself and the bees. Of course, we would think him mad. But for some reason, we see no problem with humans wanting to satisfy their emotional needs through a hypothetical connection with aliens? Why do we give this attitude our blessing?

It is strange and weird to want to have our emotional and spiritual needs filled by real aliens, and it truly makes very little sense. For starters, it is goofy for us to believe that real aliens would even be similar enough to us to provide such an emotional connection given our lineage within the simian order on this one specific planet. We can easily imagine just how different aliens could really be from us. For instance, it is certainly not outside the realm of possibility that there could be planets where Darwinian principles of survival are non-existent. Aliens from these types of planets could be very difficult for us to relate to. Or there could be planets where beings have developed with a collective sense of what we might seem to us to be telepathy and have never had a need to harm one another. We also have absolutely no idea how intelligent aliens might deal with matters such as culture, education, music, entertainment, financial systems, social structures, and the like. While these institutions appear universally inevitable and important—to the point, even, where many humans think that earth is still at the center of grand, cosmic dramas—this is not necessarily so.

For that matter, we have no idea if the factors of ego consciousness, territorialism, and Euclidean spatial awareness that have played such a key role in the human drama, are universal or idiopathic in nature. Changes to the specifics of these variables would certainly have effects on the features of an alien consciousness. E.O. Wilson has observed that our social interactions play a critical and determinative role in the development of human consciousness itself. In *The Meaning of Human Existence*, he stated,

> Until about three million years ago the ancestors of *Homo sapiens* were mostly vegetarians, most likely wandering in groups from site to site where fruit, tubers, and other vegetable food could be harvested. Their brains were only slightly larger than those of modern chimpanzees. By no later

than half a million years ago, however, groups of the ancestral species *Homo erectus* were maintaining campsites with controlled fire—the equivalent of nests—from which they foraged and returned with food, including a substantial portion of meat. Their brain size had increased to mid-sized between that of chimpanzees and modern *Homo sapiens*. The trend appears to have begun one million to two million years previously, when the earlier prehuman ancestor *Homo habilis* turned increasingly to meat in its diet. With groups crowded together at a single site, and an advantage added by cooperative nest building and hunting social intelligence grew, along with centers of memory and reasoning in the prefrontal cortex.

Conventional evolutionary biology posits a relationship between emotion, human social development, and the development of memory and reasoning. But how would the development of these higher order mental faculties change if the underlying social structures changed? We know that on planet earth alone there at least 20 different species that form eusocial organisms. Members of a eusocial organism cooperatively rear their young across multiple generations and divide labor in a manner that gives the group certain macro characteristics of organization and activity that cannot be possessed by any single member acting on its own. Eusocial activity is known to occur among a diverse, but limited, range of species on earth. These include insects, marine crustaceans, subterranean rodents, and *homo sapiens*. The fact that this activity occurs among such a diverse and dissimilar range of species on earth makes it implausible to rule out the occurrence of this form in any particular way on an alien planet. Thus, we have no way of really knowing the kinds of alien eusocial behavior that the rest of the universe can accommodate that in turn form specific types of alien minds.

Given the fact that key factors which we know have given rise to human consciousness include social organization and interaction, and given the fact that there would be no reliable way to predict a limit to the forms of social organization on alien planets, it becomes extremely difficult to determine the precise ways in which the particular form of scientific consciousness that we see among humans is truly universal, or actually idiopathic. Yet, conventional scientific thought routinely attempts to make certain predictions about the universal characteristics of alien scientific consciousness based on such limited sources of knowledge. In *The Meaning of Human Existence*, for instance, E.O. Wilson argues that the key features of any alien scientific being would include the following: (1) that the aliens are fundamentally land-dwellers, and not aquatic; (2) that the aliens are relatively large animals; (3) that aliens are biologically audiovisual; (4) that aliens have heads which are distinct, big, and located up front; (5) that aliens possess light to moderate jaws and teeth; (6) that they have very high social intelligence; (7) that they have a small number of free locomotory appendages, with particular characteristics, including at least one pair which terminates in digits; and (8) that aliens are moral. While this is certainly an accurate abstraction of the characteristics that we possess, and while it is probably true that at least some aliens in the universe would have these characteristics, this abstraction from terrestrial sources cannot be said to give us any definitive information about the characteristics of actual aliens within a real distribution that we might contact, or be said to impose universal limitations on the possible. We are just one type of alien, on one type of planet that developed within an extremely miniscule segment of cosmological time. It would be arbitrary to place such great confidence in the universality of the features of our species as we see them today. Thus, while these abstractions provide a preliminary position for thinking about certain universal characteristics, the reliability of these abstractions are only a starting point, not an endpoint, to the discussion. We have to see these issues much more clearly if we want to have a real chance and forming real alien paradigms in the future, and not just spend all of our time tripping out on aliens.

5.3.6 Conclusion

The various emotional and cognitive reactions mentioned above are factors that must be taken into account, on both a personal and collective level, if alien anomalies are to be approached as genuine objects of pre-scientific thought. Now, to be clear, alien anomalies do not have to be approached as objects of science; if we want to, we can continue to treat them as objects of fantasy, mysticism, religion, rumor, and so forth. Each individual gets to choose their own adventure here. Just as in Copernicus' day, the universe did not force humans to reckon with the actual position of the earth in space, we today do not have to approach alien anomalies in a realistic way. If we wanted to, we could continue to engage in science fiction fantasy or metaphysical, spiritual, or religious speculations for the entire duration of our lives; or, we could simply ignore them and never learn how to participate in the galactic order. Either way, this is a choice we must make today, just as Copernicus and his allies made a similar choice in their time.

But for those who are interested in generating real knowledge, it is self-evident that scientific process must be applied here because all the other processes that have been employed have led to no new knowledge in over 60 years (a state of affairs that would be unacceptable in any other domain of knowledge). In order to do that, these strange side effects of alien anomalies must be treated and resolved. For reasons that are not clear, there is something about the possibility of alien contact that seems to drive humans a little mad—that's just how things have started out here. But it doesn't always have to be that way, and, in fact, it can't always be that way if we want to start doing something real and normal here.

There are certainly many ways in which we could be interested in alien anomalies without getting too worked up about them. Think about Von Frisch again, but this time the normal Von Frisch, not the one tripping on LSD the whole time. The real Von Frisch spent decades studying bees. He didn't need to believe the bees were giving him special messages or mystical insights in order to do so. He appreciated the knowledge he was able to derive from the methods and pathways available to him. We could do the same thing with alien anomalies as well, even if every fiber in our being is telling us to relate to them purely as mysterious, supernatural beings and go crazy with them. But why not just leave the spiritual stuff to God, get our emotional needs met by humans and our pets, and keep the aliens out of it?

Indeed, it is the mastery of this kind of attitude which will be critical and necessary if we want to start gaining real knowledge here. Of course, many will never want real knowledge, and will be very content to simply relate to these beings in a mystical, religious, and magical way. But they will not generate any new content beyond what has been generated thus far and will simply spin their wheels until they die. Sooner or later, however, enough humans will quit trying to relate to these anomalies as supernatural beings and just start treating them as other types of beings we can study, exotic inhabitants of the universe that we get to examine and learn about if we're lucky enough and prepared enough to catch the opportunity when it presents itself. If Von Frisch can get into bees without tripping on LSD, certainly we can get into this stuff without needing to make it more exciting than it already is. While it may be somewhat unsatisfying at first to give up the idea that aliens are going to be our galactic friends and spirit guides, the value of the new forms of knowledge that will be revealed by a more "normal" inquiry will more than make up for the release of these childish and primitive attitudes.

5.4 On *the formation of alien paradigms*

While it would be counterproductive to try to predict what potential alien beings might be like, it would be instructive to learn what we can about any alien anomalies which actually come with our range of experience. And although hardly any scientific study of these alien anomalies has commenced, the possible human to alien encounters of which we are aware points to some rather intriguing possibilities.

Let us go back to the tic tac encounter yet again. If we direct our attention to the facts and circumstances of this situation, take it as a real occurrence, and apply a reasoned analysis to the circumstances, we can see traces of new knowledge emerging from its presentation.

Assume for the sake of argument, that inside the tic tacs there were actual biological beings maneuvering the objects and making conscious decisions about what to do and where to go.[67] If that was the case, then we can also assume that while the group of tic tacs was making its initial movements (ascending up to around 80,000 feet and descending to sea level almost instantaneously for several days) they were aware that the crew on the *USS Princeton* was watching them. We can feel comfortable making this assumption because the tic tac later appeared at the hold point, demonstrating that it was certainly aware of the content of the communications between and among the pilots and seaman in a manner that seemed extremely intentional (perhaps even

[67] If we return to the discussion of beavers, genetics, and dam building behavior in **Section 2.2**, we must concede that these alien motions cannot be the product of genetically determined behavior, but rather is a manifestation of an intelligence that far surpasses our ability to even understand it. Thus, it is reasonable for us to make "higher level" assumptions about their behaviors and motives.

“telepathic”). We can further assume that when the tic tac encounter occurred, the occupants were aware of the pilots coming into their direction and allowed it to occur. After the engagement between Fravor and the tic tac, it then disappeared into the distance, only to reappear on the hold point. If we assume that the tic tac's arrival at the hold point was intentional, then we must interpret this action as a display of knowledge, an act of non-verbal communication, similar to the honeyguide birds that show humans where beehives are located. If this was the case, then what was being communicated, what was being displayed?

Consider the possible epistemological implications of this alien gesture. Our universally accepted and unquestioned model of information generation and transmission says that information gathered or generated by one sentient being can only be transmitted to another sentient being through continuous linear casual mechanisms that rely on physical matter as the medium of communication (i.e., “sounds” through the air, “words” on paper, “bits” on computers). We have no conscious experiences of transmitting information in any other way (i.e., by non-physical means such as “telepathy” which is not considered to be a real method of transfer of information from the conventional perspective). In the case of the hold point, the coordinates for the point were generated by either a pilot or a seaman on one of the ships. The information of the coordinates was then entered into a closed encrypted system and transmitted by electronic means between the pilots and the seamen. Conventional understanding says that only those who had direct access to the signal, or the ability to directly intercept the signal and decode it, would be able to know the information of the coordinates. Yet, the tic tac could be showing us that there exists some other dimension of which we are unaware in which that information can be seen and interpreted; or, conversely, it could be showing us a form of technology that we are unaware of. In any event, no known paradigm of information transmission can easily account for such a display, thus it is certainly showing us something new, opening the mind to a new range of possibilities with respect to our understanding of information.[68] This is something we can see regardless of whether we can directly communicate with the aliens in the tic tacs.

---

[68] A similar type of thing has been observed on multiple occasions when the status of nuclear missile sites has been inexplicably interrupted in connection with the appearance of alien anomalies.

Similarly, consider the implications of the observed movement of the tic tac as it relates to our absolute notions of friction and resistance. We naturally assume that our experiences of friction and resistance are universal in nature and apply to any physical body in any physical space. Yet, somehow, the tic tac was able to move through space in a manner that bypassed what we would assume are universal limitations on the conditions of motion. This observed motion is simply "not possible" under the laws of physics as we presently understand them in the same way that the Galilean motion of the pendulum could not be understood by the fully committed Aristotelian. Moreover, the motion of the tic tac flagrantly violates deeply rooted and time-tested experiences of friction and resistance between physical objects and the conditions of their environment, which we have never seen violated, and believe cannot be violated; yet, indeed, they were violated. As we noted in **Section 2.11.1**, the rates of acceleration the tic tac displayed are absolutely impossible and appear like science fiction fantasy. Is this all that different from the type of reaction that the findings of Copernicus would have had on a regular person in the 16th century?

At a minimum, this tells us that our understanding, *as well as our embodied experience*, of these physical principles is incomplete and that somehow, both the tic tac and the fighter jets that were interacting with it are real physical objects in real physical space yet not both subject to the same sets of real limiting conditions, thus implying the existence of a "deeper" kind of space of which we are unaware. This challenges the basic notion of "physicality" itself in a profound and unsettling way. For us, the laws of resistance and friction are inescapable; for the tic tac, they are not. We have absolutely no idea how this is possible. The conventional mind may choose to ignore or turn away from this, just as Copernicus' detractors did with respect to his work. But this display can be interpreted as a signal that our paradigms are limited and incomplete in ways we cannot imagine. If we are willing to assume the anomaly was real and actual, then we open ourselves to consider that perhaps our views of physics are not as universal as we thought they were. Perhaps we are more like the Aristotelian who thinks he understands the pendulum's motion than we ought to be.

Obviously, one could go through other examples of alleged alien anomalies, assume for the sake of argument that they actually happened, and draw similar types of conclusions. Unfortunately, however, such exercises in thought will remain largely speculative until we are prepared to gather more precise and detailed data on the motions of these alien anomalies and how these anomalies are interacting with and effecting their immediate environment (such measurements must occur while the anomalies are active, and not just after they have passed). As evidenced by the example of the tic tac, the technological means for a much more detailed study of these alien anomalies is at hand. Being ready and prepared to gather the most robust detailed and precise data from the next type of "tic tac" encounter would be a valid avenue of normal scientific activity.

In light of the previous discussion in this work, we are now in a position to visualize a potential pathway towards the resolution of the type of alien anomaly presented by the tic tac. First, we need more periods of active observation of these anomalies over a period of stable time. Perhaps this process would take several decades where individuals continue to observe, record, and detect these anomalies with devices and measuring techniques that have been designed in advance for such a purpose, instead of just getting information from smartphones or other devices which were randomly at hand. One of the barriers here, of course, is the sporadic and irregular appearance of these anomalies. At present, we are unable to predict when and how they will appear (and, indeed, many believe it would be impossible ever to have such knowledge). But this problem appears primarily technical in nature. If we put as much time and resources into a scientific survey of alien anomalies as we put into generating "alien entertainment" (which amounts to billions of dollars per year), it seems like we might be able to make some headway on this issue.

Just as the alchemists and pre-scientific chemists continued to develop knowledge of the various chemical elements without really having a clear notion as to what was going on until Mendeleyev's work, so too can researchers continue to develop a new kind of evidence about these alien anomalies based on their active observation, instead of simply being obsessed with sightings from the past (which is what the overwhelming majority of all investigations into this subject matter consist of). All one really needs here is someone—or a group of someones—to play a role in relation to alien anomalies, that Mendeleyev played in relation to chemistry.

Another valid avenue, which is certainly far more controversial, would constitute a forensic study of alleged human-alien encounters. Many scientists would resist the idea of giving any type of weight to alleged human-alien encounters in the absence of direct scientific proof confirming the reality of a given encounter, but if we are in a pre-science phase of discovery then we should admit that reports of these encounters have merit as a prelude to normal scientific activity. Of course, this is not to say that anecdotal reports without corroborating physical data must be taken as real, but simply that gathering information from anecdotal reports could have some value in shaping future research activities.

In this connection, if one accepts the analogy between the Copernican Revolution and a potential scientific revolution involving alien anomalies as valid, then one would also see the importance of transforming the influence that institutions may have on scientific discourse. In the days of Galileo, the primary institution was the Church. Today's version of the "Church" in relation to alien anomalies is what we could refer to as the "National Security State", which has been painstakingly chronicled in the scholarship of Richard Dolan and others. While we could engage in reasoned analysis about the reasons why the National Security State has decided to suppress alien anomalies, and the means which it has employed in so doing, the important point is simply to acknowledge that this institution has had a suppressive effect on the development of science which needs to be changed.

In the 17th and 18th centuries, the navigation of the seas was greatly hampered due to the fact that humans had no way to accurately calculate longitude. This inability to determine longitude was very costly in terms of resources and human lives. For a long time, many thought it was even impossible to do and thought that the ability to accurately reckon longitude would always remain "unknown". Eventually, public clamor forced the government of Great Britain to do something different. In May of 1714, a petition by the merchant and seaman community of England arrived at Westminster Palace and was taken up by a Parliamentary Committee. The Parliament sought the counsel of Sir Isaac Newton. Newton prepared an analysis of the problems inherent to determining longitude to the committee and discussed different approaches to resolving the problem. Parliament considered Newton's analysis and determined that the best approach was to solicit solutions to the problem from any source and provide a reward to anyone who could solve the problem.

As a result, the English government passed the Longitude Act of 1714. The Act offered 1st, 2nd, and 3rd place prizes. The 1st Place prize of 20,000 pounds (which would be equivalent to $ 1 million dollars today) went to the person who could provide a reliable means to calculate longitude within an accuracy of half a degree. In addition to offering the prizes, the Longitude Act also established what Dava Sobel refers to in her book, *Longitude: The True Story of a Lone Genius Who Solved the Greatest Scientific Problem of His Time* (1995), as a "blue ribbon panel of judges that became known as the Board of Longitude". Sobel describes this Board as follows,

> This board, which consisted of scientists, naval officers, and governmental officials, exercised sole discretion over the distribution of the prize money. The astronomer royal served as an ex-officio member, as did the president of the Royal Society, the first lord of the Admiralty, the speaker of the House of Commons, the first commissioner of the Navy, and the Savilian, Lucasian, and Plumian professors of mathematics at Oxford and Cambridge Universities.
>
> The board, according to the Longitude Act, could give incentive awards to help impoverished inventors bring promising ideas to fruition. This power over purse strings made the Board of Longitude perhaps the world's first official research-and-development agency. . . In order for the commissioners of longitude to judge the actual accuracy of any proposal, the technique had to be testified on one of Her Majesty's ships, as it sailed "over the ocean, from Great Britain to any such Port in the West Indies as those Commissioners Choose...without losing their Longitude beyond the limits before mentioned."

Obviously, the gambit worked. An "alien anomalies" version of the Longitude Act of 1714 would be extremely helpful in moving scientific process forward. The means to understand the underlying causes of these alien anomalies are clearly at hand, as evidenced by the small amount of data we have been able to gather from anomalous encounters involving these strange objects and military service members in the early part of the 21st century. These encounters clearly indicate the concrete pathways we could pursue for further study of these matters. If these alien anomalies have prosaic explanations, science should be able to demonstrate as such and close the book on them once and for all so that everyone can move on. On the other hand, if science confirms that the cause of at least some of these alien anomalies are in fact "real" aliens from other places, then it opens the door for a deeper exploration and understanding of this phenomenon which could be just as revolutionary for the modern era as Copernican Revolution was in its time.

5.5 *The translation of disclosure*

In general, Kuhn's work counsels us to abandon the current obsession with disclosure as an act of revelation, and simply work on engaging in scientific practices that resemble normal scientific activity. This, of course, does not preclude us from continuing to insist that the U.S. military disclose the information that it has compiled (because obviously, such information would be helpful here), but it does counsel is to stop putting all our eggs in one basket, so to speak, and thinking that only through disclosure will we gain real knowledge.

The record of history tells us that if we set a scientific process in motion, one day there will either be alien paradigms within which alien anomalies are no longer anomalies, or we will determine that aliens are akin to the ether—something that scientists were sure in the late 19th century was real, but led them to discover relativity instead of what they thought they were looking for. In any event, somehow, in ways that we cannot yet predict, we will one day find ourselves in a foreign universe of new understanding. Thus, we conclude this section with the following remark on "translation" from Kuhn,

> For most people translation is a threatening process, and it is entirely foreign to normal science. Counter-arguments are, in any case, always available, and no rules prescribe how the balance must be struck. Nevertheless, as argument piles on argument and as challenge after challenge is successfully met, only blind stubbornness can at the end account for continued resistance.

That being the case, a second aspect of translation, long familiar to both historians and linguists, becomes crucially important. To translate a theory or worldview into one's own language is not to make it one's own. For that one must go native, discover that one is thinking and working in, not simply translating out of, language that was previously foreign. That transition is not, however, one that an individual may make or refrain from making by deliberation and choice, however good his reasons for wishing to do so. Instead, at some point, in the process of learning to translate, he finds that the transition has occurred, that he has slipped into the new language without a decision having been made. Or else, like many of those who first encountered, say, relativity or quantum mechanics in their middle years, he finds himself fully persuaded of the new view but nevertheless unable to internalize it and be at home in the world it helps to shape. Intellectually such a man has made his choice, but the conversion required if it is to be effective eludes him. He many use his new theory nonetheless, but he will do so as a foreigner in a foreign environment, an alternative available to him only because there are natives already there. His work is parasitic on theirs, for he lacks the constellation of mental sets which future members of the community will acquire through education.

The conversion experience that I have likened to a gestalt switch remains, therefore, at the heart of the revolutionary process. Good reasons for choice provide motives for conversion and a climate in which it is more likely to occur. Translation may, in addition, provide points of entry for the neural programming that, however instructible at this time, must underlie conversion. But neither good reasons nor translation constitute conversion, and it is that process we must explicate in order to understand an essential sort of scientific change.

6.0 ALIEN EXPERIENCES

An alien experience is a human encounter with an alien anomaly. There are two general types of alien experiences: the personal and the impersonal.

Impersonal experiences constitute situations in which the alien anomaly is observed but does not appear to have an intimate effect on the observer. Scientific study is one example of this class of experience. Additional examples include situations where a person sees an alien anomaly, can't explain it, but otherwise goes on about their lives as normal. The alien experience might be interesting to them, and something that is remembered fondly, or forgotten, or even blocked out. Either way, the impersonal experience does not seem to change the individual's personality structure or the reality of their life in a noticeable way. *See* **Section 2.9.** In these cases, there is a safe observational distance between the alien anomaly and the human reality of the observer.

Personal alien experiences are different. In this class of experience, the alien anomaly does seem to have a direct and intimate impact on the person and changes them in one way or another. The safe distance between the observer and the alien anomaly is dismantled once and for all and nothing is ever the same again. Sometimes these changes are positive and result in a new appreciation for the depth of cosmological time and space and the alien lifeforms it can produce. Such feelings can then lead to productive forms of exploration and integration. But in other cases, the personal alien experiences can be destructive, leading either to idiopathic neuroses, such as chronic estrangement from the world, or various pathological conditions such as a "messiah complex". In common parlance, the term "close encounter" is another way to describe this class of alien experience.

6.1 The intrinsic strangeness of alien experiences

In the modern world, alien experiences are intrinsically strange. On the one hand, there appears to be a rational argument for the existence of intelligent alien beings throughout the universe, and it is hard to understand why aliens shouldn't be visiting earth already—shouldn't they be all over the place by now? ("Fermi's paradox"). On the other hand, the possibilities of aliens and humans actually communicating and encountering each other on planet earth appears insane and absurd on its face, like something out of a science fiction movie.

But this sense of absurdity or implausibility is not necessarily a sound argument against the reality of such occurrences. Imagine, if you will, a time-travelling 21st century molecular biologist. Imagine that this biologist, armed with a presentation on genes, germs, viruses, and various microbes, was able to travel back to the mid 16th century and give a lecture to the denizens of Nuremberg or Basil. Would these people find her presentation any less alien—or more believable—than the events depicted in the Nuremberg or Basel broadsheets? To an informed citizen of the mid 16th century, modern-day molecular biology would be insane, absurd, and unbelievable—and rightfully so. We can see that much by considering the limits of knowledge in those days as inferred from the written record of human history. We know that a person in the 16th century could not possibly understand modern biology, as the conceptual structures which permit these facts were built up over centuries.[69] We can also understand that it would be accurate and rational for a 16th century human to find the ideas of modern biology absurd and insane from the point of view of the paradigms that were conventional and stable for them. At the same time, we can also see that a rational 16th century resistance to modern ideas of biology is not valid or correct within 21st century systems of thought, to the perspective in which we operate. It is only the relativity of the paradigms that creates the sense of strangeness and not necessarily the Truth itself.

The intrinsic strangeness of alien experiences may come down to this: on the one hand, we see no rational reason why aliens shouldn't already be visiting earth; on the other hand, when a human claims they have encountered an alien, their reports sound like fantasy science fiction stories and our minds instinctively find them impossible to believe.

6.2 Modern alien experiences

---

[69] Think back to the discussion presented in **Section 5.2.2** with reference to Paracelsus and the way people viewed the world in those times. It would literally be impossible for such people to understand modern science through no fault of their own.

Ezekiel's encounter with the chariot is an ancient example of an alien experience. While the encounter could have been caused by a real alien, at a minimum the experience was certainly unfamiliar, strange, suspicious, other, and therefore "alien" in at least the prosaic sense of the term. Ezekiel's experience appears to have been integrated and gained the imprimatur of social legitimacy, as shown by the fact that a record of his experience has been passed down through generations as a well-known part of one of the world's major religious traditions.

Alien experiences appear to have always existed alongside those constituting the normal, everyday experiences of regular humans. In *Passport to Magonia* (1969), for instance, Jacques Vallee provides an extensive catalogue of the "reports" of such experiences going back centuries. In the modern era, personal stories of alien experiences began to emerge in the 1960's and 1970's as a particular genre of literature. In 1965, Vallee wrote a book entitled *Anatomy of a Phenomena: UFOs in Space*, which provided a modern orientation to the subject. Vallee followed this initial publication with a string of others including *Passport to Magonia, Invisible College* (1974). Since his initial work in 1965, Vallee has written 20 more books, all variations on the same themes he's started with. Vallee has addressed audiences all over the world, consulted with high-ranking government officials, and appeared in countless documentaries. Vallee has achieved iconic status as a figure within the modern-day UFO movement and has made a positive contribution to its popularity. Vallee was supposedly the inspiration for the main character in Stephen Spielberg's movie *Close Encounters of the Third Kind* (1977).

In 1987 a writer by the name of Whitley Streiber published a book called *Communion*, in which he described what he alleged were personal encounters with alien beings. From 1987 until the present, Strieber has served as one of the lead spokespersons for those who have had alien experiences. These people often colloquially refer to themselves as "experiencers". After the publication of his work in *Communion*, Strieber began receiving a large volume of correspondence from readers who also claimed to have alien experiences. Some of these letters were published in a book called *The Communion Letters* (1997). Like Vallee, Strieber remained a prodigious writer and spokesperson in the realm of alien experiences and has continued to write and speak to this day (2025). Streiber represents a class of experiencers that is growing every day, both in terms of the content they generate and the types of effects they have on the social order itself.

Strieber and Vallee represent two distinct poles in the realm of modern alien experiences. Vallee plays the role of an interested scientist, someone who takes the subject matter seriously and wishes to understand it. Vallee is open to any and all theories about the causes and meaning of alien anomalies and engages in speculation to arrive at certain theories. Vallee is willing to give the benefit of a doubt to nearly all accounts of alien experiences and willing to let all the alien evidence in, with little regard to its admissibility.

Strieber shares Vallee's openness and approach to alien anomalies, but Strieber also speaks from his personal alien experiences. In general terms, one can describe the nature of his experiences as follows. Strieber was more or less a normal person for most of his life, a man interested in what one might call new age philosophy, but no different than many modern-day urban humans. Strieber was also an accomplished writer and a family man, until one December night in 1985 when he claimed to be visited by beings he called throughout his life "the Visitors". From that point on, his interior life radically changed and became a personal journey through a strange new world which continues to this day in the year 2025. His alien experience called him to question everything about his past, his present, his future, the nature of reality itself. Regardless of whether one believes that what happened to Streiber was real or not,[70] Strieber's experience fundamentally altered his world. Strieber's work describes, in painstaking and personal detail, one person's journey through a strange, alien world that exists alongside the normal, everyday world, but also seems to transform it.

---

[70] Strieber himself often questions whether his experiences are real or not; thus, it would be unfair to accuse Streiber of being insane or deluded, even if one ultimately finds his work unconvincing, absurd, or unbelievable. Moreover, Streiber maintained normal attachments throughout his life, including a successful marriage, fatherhood, and a flourishing career. The author of this work had an occasion to hear him speak and meet him briefly and found him to be more or less a regular guy. Thus, as crazy and outlandish as Strieber's claims or experiences might appear, he nevertheless maintained what can also be called a regular, "normal" life from the conventional point of view.

While the work of Vallee and Strieber is important in generating interest and respect for the subject matter and has value from the standpoint of human experience in general, their work does not seem to reveal any real terrain with respect to alien anomalies beyond that previously mapped by Jung in 1958. They have certainly fleshed out more details and provided more color, but they have not exceeded the parameters of the initial Jungian discourse. No mysteries have been answered, and no greater understanding had been developed. The aliens in these works still remain objects of mysticism, metaphysics, and spirituality.

Similarly, the work of academics in the field of alien experiences don't seem to have surpassed Jung's initial cartography. For instance, D.W. Pasulka, a professor of religion at North Carolina University, has written two books: *American Cosmic* (2019) and *Encounters* (2024). Another academic, Dr. James Madden, a professor of religion at Benedictine College in Iowa, has published a book entitled *Unidentified Flying Hyperobject* (2023) which also seems to cover the same ground.

The examples of Vallee, Strieber, Pasulka, and Madden, represent the tip of the proverbial iceberg when it comes to the subject of alien experiences. There are numerous books, documentaries, and active social media channels that explore veins that are very similar to the work of these individuals. But they all generally speak to the same common themes; alien experiences are weird and difficult to reconcile with the normal world. They are difficult to integrate with more ordinary forms of experience. The mind naturally is led to relate to them as mystical, religious, or spiritual experiences.

6.3 *The value of personal alien experiences*

Some of the claimed modern alien experiences are so insane and outlandish that the mind wants to immediately dismiss them as false and delusional. However, it would be foolish to reject a serious consideration of personal alien experiences for several reasons. First, regardless of whether they "really happened" or not, these experiences speak a human truth about our present relationship to alien anomalies. In the case where the alien experience seems completely real to the individual, a real and genuine human reaction will occur. As noted above in **Section 2.9**, when the individual believes the experience is real—even if it is not—it will still have a real effect. So even if someone who is not "in" the alien experience itself is not convinced the alien anomaly was real, if the person having the alien experience is so convinced, then a human truth will be revealed. And, as noted in the references to Jung in **Section 5.3.1**, alien experiences teach us something about us—our hopes, fears, and inclinations—if nothing else.

At a minimum, alien experiences reveal our subconscious reactions and attitudes towards alien anomalies, bringing them to the surface, into consciousness, where we can examine them and discuss them. There may be a benefit in such a collective, deliberative analysis. While we may not be able to objectively verify the reality of alien experiences at present by correlating them with physical phenomena due to the lack of a sufficient observational regime, and are forced to take these experiences on faith, alien experiences may still express a human truth from the standpoint of religious, spiritual, or metaphysical thought and thus have value and meaning in that kind of way.

Second, we should be cognizant of the fact that our science fiction fantasy conditioning has entrained us to believe that experiences which sound like science fiction fantasy, must indeed be fictional. Thus, when an individual describes alien experiences that sounds like something we might see in a movie, our initial inclination is to reject the experience as fictional in nature and not real. But, turning again to the example of the Copernican revision, we should be wary of these initial impulses and take a more open attitude. In the days of Copernicus, most rejected the paradigm presented by Copernicus because the implications of the new paradigm seemed insane to them as they clashed with the truth they knew from everyday experience. But the clash arose not because the Copernican paradigm was wrong, but because there were no Newtonian and Galilean paradigms that described things like gravity, acceleration, velocity, and centrifugal force. These would be alien paradigms to the people in Copernicus' time and would only arise at a later date. Applying Nimoy's Dictum, we are forced to admit that it is possible that a real alien experience that may sounds like a science fiction movie could actually occur, and would, in fact, appear fantastical in that way. In other words, the fact that some alien experiences may sound like something out of a science fiction movie is a poor reason to reject them as possibly real. While there may be other good reasons to reject them as being real, their similarity to science fiction fantasy is not one of them.

6.4 *The Pascagoula Encounter*

There are so many alien experiences in the modern reports that one could spend a lifetime studying them and never come close to exhausting their supply. However, one reported experience stands out as particularly instructive in light of the type of discussion presented in this work. This is an alleged encounter which occurred in Pascagoula, Mississippi on the night of October 11, 1973, involving two regular men who claimed they were briefly taken aboard a spacecraft and examined by alien beings.[71] While this case lacks any tangible direct evidence of an alien anomaly, the circumstantial evidence from this encounter is quite strong and compelling and it provides a fascinating case study of an alien experience.

6.4.1. The weight and meaning of circumstantial evidence

Before considering this case, a word about the relative weight given to judgments formed based primarily on circumstantial evidence is in order. As we have seen from the discussion of Kuhn's paradigm theory, there is a period of immature pre-science which may precede the ripening of normal scientific processes and procedures in some instances. This less mature phase is characterized by evidence that is less rigorous and more circumstantial in nature. *See* **Section 4.1.** But such evidence has real value in the overall context of scientific process and provides a meaningful platform for later activities. If this evidence is rejected outright because of its poor quality, and inquiry is shut down, nothing new is learned. By contrast, if the poor quality of the information or data is taken as an invitation to develop better means of measurement and detection, new forms of knowledge may possibly emerge.

---

[71] There are a wide range of sources one could consult on this particular encounter, as it was reported widely and investigated by numerous people. Here, we are relying primarily on two sources, *UFO Contact at Pascagoula* (1983) and *Pascagoula: The Story Continues* (2019). One can also find a wide range of material regarding this encounter on the internet, including interviews and some of the original news footage by local reporters in the aftermath of the encounter. The interested reader is encouraged to "do your own research", as they say.

The modern-day world of conventional science does not have direct experience with the kind of pre-scientific activities that people like Bacon, Boyle, and even Mendeleyev had to contend with. Today's conventional scientific world is driven by institutions, peer-reviewed journals, and well-established systems of conventional pedagogy and initiation. Even if the results of modern research are sometimes unconventional and strange (as in the case of subatomic physics or cosmology), the underlying platforms are not because they are still based on well-established means of measurement and observation. There is a high degree of conventional support in modern science which should not, by definition, exist in a pre-scientific study of alien anomalies; thus, the study of alien anomalies lacks conventional supports, presents poor types of data at present, and is subject to easy rejection by the weight and inertia of conventional systems of thought. However, the Kuhnian perspective places the status of this new, alien field in a broader historical context within the human record, allowing us to see why this pre-scientific phase actually makes good sense. For a real study of alien anomalies, if it is to actually occur, would only happen outside of the conventional structures, within the messy, inconclusive, and puzzling facts of the real world since it is so alien to the conventional human narrative developed thus far.

In this context, real evidence in the realm of alien anomalies will necessarily begin with evidence that is circumstantial and anecdotal in nature. Such evidence should not be rejected because it does not meet the rigorous requirements of the laboratory or the peer-reviewed journal, but its weight should be adjusted within the overall context of normal scientific process. While conventional understanding is correct to conclude that the weight derived from circumstantial and anecdotal evidence is insufficient to establish scientific facts, conventional understanding errs when it decides to shut down further inquiry into the matter unless better proof is presented. Instead, the correct approach is to examine the evidence for what it's worth and not ascribe to it any greater weight or value than it is entitled to.

In the realm of legal judgments, circumstantial evidence is sufficient to convict a person of the most serious of criminal offenses and deprive them of their liberty. This is a signal that circumstantial evidence is sufficient for human observers to establish meaningful facts about things that have happened in human reality (even if it may be insufficient to establish what we would consider to be "scientific facts", or, more specifically here, facts about objects from one of the invisible orders of the world, which is the order in which we must place alien anomalies at the present time). *Consider, e.g.*, **Sections 2.4** and **2.9** (making the argument that scientific process always aims to move inferred objects from invisible, unknown orders, into those in conventional reality). The power of circumstantial evidence derives primarily from the inferences one draws from an analysis of known circumstances based on principles of common sense and reason. In such cases, the outside observer is denied direct access to the event itself but can observe and analyze the aftereffects of the event through the responses and behaviors of other humans who were involved in it, consider the evidence rationally, and come to certain conclusions about what most likely happened. To put this in another way, if one analyzes the patterns of fingerprints, one might be able to infer what a hand was doing at the time it acted, even though one was not present to witness the actual action of the hand itself.

Thus, in examining the Pascagoula Encounter, several things must be borne in mind up front. First, there is little to no direct proof or evidence in this case. There are no photographs, no physical traces, no radar readouts, no types of measurements at all. Instead, all we have are reports of experiences by various individuals, situated in different social positions in relation to the encounter. In considering the Pascagoula Encounter, the best we can do is to determine whether or not something real likely happened, or whether the reports were the product of delusion, intentional deception, or mistake. We cannot establish anything beyond that given the nature of the evidence at hand. Thus, anything we learn from an examination of this encounter will be quite limited and perhaps unsatisfying.

Nevertheless, an analysis of this encounter is useful for two reasons: 1) it gives us a fascinating and instructive example of how alien experiences can effect different individuals; and 2) it reemphasizes the importance of developing normal scientific practices and procedures that can be deployed in a way to capture evidence and information about these types of encounters in real time, instead of only dissecting the ashes of these events after they have transpired. With these preliminary considerations in mind, we will now turn to a discussion of the circumstances of encounter itself and its aftereffects.

6.4.2 A description of the Pascagoula encounter

At the center of the Pascagoula Encounter are two individuals, Charlie Hickson and Calvin Parker. At the time of the encounter, Charlie and Calvin were both workers at a local shipyard in Pascagoula. Calvin was 19 years old, Charlie was 42. Charlie was a veteran of the Korean war and had been involved in five major wartime battles, where dying was a distinct possibility for him. Calvin was a friend of Charlie's son. By all accounts, Calvin was a normal 19 year-old at the time. Calvin and Charlie both worked at a local shipyard in Pascagoula, with Charlie in somewhat of a supervisory position at the shipyard, and Calvin working as a welder.

On the evening of October 11, 1973, Charlie and Calvin decided to go fishing after work, which was a favorite pastime of the two of them. On that date, they were fishing at an old shipyard as the evening turned to night, when they heard a zipping sound and saw what appeared to be a flying craft about 60 to 70 feet away from their position. The men claimed to be completely surprised by its appearance. They heard no engine noise but saw only a blue, pulsing light, as the craft approached closer to them, hovering silently above the ground. According to Charlie, it was approximately thirty feet long, eight to ten feet high, and shaped like a football.

As the craft hovered before them, they were uncertain as to what do to or what would happen next, when an opening appeared in the craft and the blue light of the craft went out. A strange glow then issued from the opening. Soon, three beings appeared from the opening and approached them. Charlie reported that he shook in fear as the beings approached, while Calvin screamed. The beings reportedly glided out of the opening above the ground and drew closer to them. Charlie described what happened next as follows:

> If they had a more human likeness it would not have shocked me so. The head seemed to come directly to the shoulders, no neck and something resembling a nose came out to a point about two inches long. On each side of the head, about where ears would be, was something similar to the nose. Directly under the nose was a slit resembling a mouth. The

> arms were something like human arms, but long in proportion to the body; the hands resembled a mitten, there was a thumb attached. The legs remained together and the feet looked something like elephant's feet. The entire body was wrinkled and had a greyish color. There could have been eyes, but the area above the nose was so wrinkled I couldn't tell.
>
> Two of them took hold of my arms, one on each side. Instantly, I felt pain in my left arm, but just as quick it was gone, I became helpless; I could not move and had no sensation of feelings. One of them took hold of Calvin; I saw him go limp and found out later he had fainted. As they were taking me aboard the craft I remembered reading on different occasions in the past where people had disappeared or vanished. "Will this happen to us? Will anyone ever know what happened if they take us away? Will this be my last day on earth?" All these things were racing through my mind.

From this point, the encounter sounds like something out of a science fiction movie. Charlie reported that once inside the craft, they were examined by the beings with "something that resembled a big eye [that] seemed to come out directly from the wall". Charlie claimed that he wanted to close his eyes during the examination, but that "some force kept them open." Charlie claimed that during the examination he felt "no pain, no sensation" and was unable to move his body. After the examination concluded, he tried to move and locate Calvin but didn't know where he was. As he stayed in this suspended state, Calvin reported the following thoughts went through his mind.

> I was alone for what seemed like an eternity, but couldn't have been more than a few minutes. My mind wandered back to a crossroad north of the 38th Parallel in North Korea. The enemy had thrown up a roadblock to cut us off from our main unit, but by being calm and keeping our heads we learned the North Koreans a lesson that day, one that many of them didn't live to remember. "Keep calm, Charlie," I kept telling myself. "Don't panic, your time will come, they can't keep you this way forever." Then I saw them again. They took hold of me the same as before. The opening appeared again and we glided toward it, "Maybe this isn't my day to die." We glided through the opening and moved towards the place they had taken me

from. My feet were touching the ground. When they released me, I fell when my legs gave way.

"My God, what have they done to him?" Calvin was standing there with his arm outstretched, almost in shock. There was more terror on Calvin's face than I have ever seen on anyone's. I started to crawl towards him, but rose to my feet before I reached him. Just before I reached him, I heard the zipping sound again and as I looked around, I saw the blue light flashing lights and almost instantly the craft was gone. I can recall something raced through my mind. "We are peaceful, we meant you no harm."

After this encounter, the men were in a state of shock. Charlie, relying on his wartime experience, was able to gather himself far more than Calvin, who seemed to remain in a state of mental and emotional disability. After debating their next course of action, the men decided to call the Keesler Air Force base near Pascagoula to report the incident. Charlie claimed they did this out of a concern that the beings would return. The Air Force base informed them that they did not handle such reports and that they would have to contact their local sheriff.

In his memoir of the encounter written in 2019, Calvin included in the appendix of his work several clippings from the local newspaper that reported on the event at the time. The following is taken from an article entitled "Pas men 'forced' aboard UFO here" from the *Mississippi Press* on October 12, 1973, describing the reports from local law enforcement officers.

Captain Glen Ryder said he didn't believe the men at first. "I thought they were pulling my leg. The young boy (Parker) was scared to death. He was almost in shock."

The officer said he tried everything he knew to break their stories, but they persisted. He said he then placed them in a cell, which had a hidden tape recorder. They didn't know about the recorder, said Ryder.[72]

"My God," Hickson said on tape. "I never seen anything like this in my life. My nerves are a wreck. I got to get out of here."

---

[72] A full transcript of this recording is produced in *UFO Contact at Pascagoula*. The recording itself is also available on the internet for listening.

Ryder said Parker at this time was praying.

"I didn't believe their story at first," said Ryder, "but now I do after I got them on tape."

"If they were lying to me, then they should be in Hollywood."

The officer said he contacted Hickson's wife, who told them that her husband sometimes would take a drink, but he never became inebriated.

They were not drunk, the officer said. He said he could smell whiskey on Hickson's breath, but the man had told him he took a drink after they were freed.

The captain said they received a report from Larry Booth, operator of Standard service station on Hwy 90 at the intersection of Market St., who told them he saw a UFO about 8 p.m. headed east towards Mobile. He told officers it had a flashing blue light. Several sightings were reported to the sheriff's department from the Pinecrest area.

A scheduled press conference at 11:15 this morning was cancelled by the two men.

According to Sheriff Fred Diamond they have decided not to make any more public statements. The sheriff said the two "are willing to talk to federal officials."

Diamond also said they would like to go to Jackson to take a lie detectors test.

"I believe they're sincere," Diamond said, "and I believe their story is true."

6.4.3. Aftermath of the encounter

After the men left the Jackson County sheriff's department, they went home, intending to return to their normal lives the next day. According to Charlie, the men had asked Sheriff Diamond to keep the information they had relayed to him confidential. But when they returned to work the next day at the local shipyard, Charlie was called to the telephone at the main office and was surprised to be receiving a long-distance call from a reporter in Jackson, Mississippi wanting details of the encounter. From that point, inquiries from reporters all over the United States overwhelmed the men, and within an instant their experiences became the subject of widespread reporting. There was no evidence or indication that the men themselves sought this public attention.[73]

---

[73] Interestingly, reports of the encounters also appeared to induce "rumor effects" in the local populace. A story from the *Mississippi Press* from

Soon, two scientists arrived in Pascagoula to conduct an investigation. Dr. James Harder, a professor at the University of California, was associated at the time with the Aerial Phenomena Research Organization. Dr. J. Allen Hynek was a professor of astronomy at Northwestern University and had been involved as a consultant to the U.S. Air Force on several studies into alien anomalies, including Project Sign, Project Grudge, and Project Blue Book. Calvin and Charlie themselves did not ask the men to visit them, but they showed up on their own initiative to investigate the reports they had heard.

In one of the appendixes to Calvin's 2019 memoir, *Pascagoula: The Story Continues*, he includes a clipping from another contemporaneous article in the *Mississippi Press* from October 14, 1973, entitled, "Scientists term Pas UFO report as 'true'". In the article, it is reported that Hynek and Harder both concluded that the men's reports were probably true. Both of the men were subjected to hypnosis by Dr. Harder, who commented upon the observations from those sessions and explained that the men truly appeared to be reporting lived experiences during the sessions.

In the article, Dr. Harder gave his conclusions as follows,

> Dr. Harder said that Parker and Hickson had faced experiences for which they were totally unprepared. He compared their experiences to that of an aborigine in Australia who is suddenly confronted by a jumbo jet.
>
> Dr. Harder said their experience "was indeed a real one. It was no hallucination."
>
> He said it would be practically impossible for the two men to simulate their feelings of terror while they were under hypnosis.

October 18, 1973, entitled "Cars jam I-10 after UFO Report" (which is contained in Calvin's book Calvin's 2019 memoir, *Pascagoula: The Story Continues*), reported that nearly "3,000 Mobile area residents blocked interstate I-10 seeking a rendezvous with flying saucers". The story goes on to report how caravans of motor vehicles flocked to sites where UFOs were thought to be gathering, creating "out of control" traffic, widespread reports among the populace, and a general sense that something real and significant was happening in the area around the sightings. This situation calls to mind those visionary rumors discussed by Jung in **Section 2.10.**

Neither of the two men were willing to be hypnotized at first, he said, but agreed after two others in the room went under to show them it wouldn't hurt them.

Hynek and Harder both vouched for the authenticity of the men's experiences, while expressing no opinions on what it was that they had actually experienced. The *Mississippi Press* article quotes Harder as saying, "They have been through a very trying experience. It's been terrifying." Hynek concurred, imploring the public not to ridicule them by saying, "They had a very real and frightening experience."

In a subsequent interview with Sheriff Diamond reported on October 18, 1973, in the *Mississippi Press* entitled "Sheriff's View", Sheriff Diamond expressed his public support for the men. In the article, Sheriff Diamond recounted in detail the investigative steps that his office took upon receiving the reports. After providing the initial description, Sheriff Diamond stated as follows,

"For several hours that night we listened to what both men had to say. We questioned them separately. Then, we placed them together in a room and monitored their conversation. Their stories—told individually and together—were the same. To them, their experience was real. They showed it emotionally.

"Our objective was to hear the men out. To listen to their complaint—in the same way we would listen and investigate any—was to check out, as best we could in a case like this, what these men said happened to them.

"In a case such as this, there is always a great deal of public speculation about what really happened. The sheriff's department has no private or public opinion about what really happened. Our position when we received the complaint was that Mr. Hickson and Mr. Parker were citizens asking for help. And, we tried to help as best we could.

"We have received reports of UFOs throughout the week. The phones have been busy 24 hours a day. I estimate that we have received more than 2,000 calls from news reporters from around the world wanting information and from people in the area who wanted to report a sighting. The calls came in from every corner of the United States.

"Throughout this situation we made every attempt to bring the federal government into the case. The Air Force, which until 1969 had the responsibility for reported UFOs, told us there was nothing they could do unless human life was in danger.

"It is certainly unfortunate that this situation occurred. Unfortunate for Mr. Hickson and Mr. Parker who underwent a dreadful experience of some kind—an experience that was very real to them. And, it has been unfortunate for our community because the situation has caused a great deal of concern."

As far as the record reveals to us, Charlie and Calvin were not ridiculed in Pascagoula, and their experiences were taken as true by local residents. Indeed, the experiences of Charlie and Calvin were validated to such an extent that on June 22, 2019, the town of Pascagoula commemorated the encounter with a historic marker at the site where the craft allegedly appeared, telling the story of what happened to the two men on that evening so long ago.

The alien experience of these men had profound and long-lasting effects on their lives in different ways. For Calvin, the effects were painful and negative, and he had great difficulty coping with the personal effects of his alien experience. Three weeks after the encounter, Calvin was hospitalized after suffering a nervous breakdown. After he rehabilitated, he attempted to join the Marines but was discharged after basic training for medical reasons. Calvin was resistant to the public attention his story generated, and generally shunned speaking about it to others, and avoided publicity. We could say that Calvin's life was haunted by the experiences of October 11, 1973, until he found some peace with his experiences, nearly 45 years later. In fact, it wasn't until Calvin wrote his first book, *Pascagoula-The Closest Encounter: My Story*, in 2018, that he felt he was finally able to process and integrate his experiences. As Calvin subsequently described in *The Story Continues*,

> I found it really quite scary thinking about writing a book and not knowing what the future might hold. The first positive thing writing the book did for me was bring me closer to my wife. The main reason for this was that now I could talk to her about what happened to me and share my feelings. I had always found this a difficult thing to do. My wife and I were now able to take trips together, work on the book together and let her

> discover things about me that she never knew before, even though we had been married all these years.

Charlie had quite a different experience, and the differences between the effects a common alien experience had on the lives of these two men speaks to the subtle and interesting ways the human psyche reacts to alien anomalies. Soon after the encounter, Charlie realized that the great public interest in the story left him no choice but to publicly speak about it to as many people as possible. In December of 1973, Charlie appeared on the *Dick Cavett Show*, discussing his experiences. Charlie had taken and passed a polygraph test, the results of which were discussed on the show. Charlie was invited to speak to different audiences at conferences and seminars and was regularly interviewed for television. Charlie was evaluated by psychiatrists and found to be of sound mind. Charlie continued to be a productive member of society and enjoyed positive and peaceful relations with friends and family. We might say that Charlie was viewed as a celebrity of sorts.

Charlie also claimed to have several additional encounters after the initial incident[74], and in each of these encounters his feelings of trauma or fear were replaced with curiosity and openness. Perhaps we can attribute this to fact that Charlie had been through traumatic wartime experiences and had learned how to keep a level head during intensely abnormal situations.

The first alleged encounter occurred in January of 1974. Charlie was hunting squirrels by himself at a local tree farm when he claims the following events transpired.

> The shotgun was lying across my lap. I had just topped off the sandwich with an orange when it dawned on me that I hadn't seen any movement at all around me—not even any birds, that seemed real strange. I saw it then, the same craft Calvin and me had seen before, about twenty five yards away in a small clearing, hovering above the ground. Before I could even think a 'radio' seemed to have come on in my mind:

---

[74] Calvin reported one encounter involving missing time which occurred one day when he was fishing by himself in 1996. When he was subsequently hypnotized, he recalled details of the experience. But he did not have the type of conscious contact experiences that Charlie reported.

"We mean you no harm. We mean no one any harm. You may communicate with us later. You have endured. You have been chosen. There is no need for fear, we will communicate again."

The 'radio' was turned off, the craft was gone, I had not moved . . . I know now I must let everyone know some way that they are up there and they mean us no harm, that possibly they can help us in the future. Since that day I have felt obliged to go to many places I have been and tell what happened to Calvin and me . . . hoping it might someway prepare the people of our world for the things that will come in the future.

In February of 1974, Charlie claimed to have another experience. This time, he claimed to be awakened one night by a dog barking outside. He went to go investigate it and "saw the dog run away as if someone was chasing him." The 'radio' inside of him was turned on again and he claimed to hear the following:

"You must tell the world we mean no harm. Your world needs help. We will help in the future before it's too late. You are not prepared to understand yet. We will return again soon."

In this incident, Charlie did not see any craft, but only claims he heard the "radio". But on Mother's Day of 1974, he had a far more dramatic encounter, this time experienced by several of his close family members as well.

Charlie, his wife, and his children, had driven from their home in Pascagoula, to Sandersville, one hundred and thirty miles north, to enjoy Mother's Day with this his parents on their farm. After spending the day there, Charlie and his family headed back home. In addition to his wife (Blanche), younger children, adult daughter (Shelia), son-in-law (Kenny), grandbaby, and Kenny's younger brother (Ernest) accompanied them as well. Charlie described what happened during this encounter as follows:

Kenny was driving slowly because of the crooked road. We were about halfway between Saucier and Ocean Springs when I noticed out the left back window that a large light, a long way off was following us. As I kept watching it began to get closer, I saw Shelia punch Kenny's younger brother and show it to him. I called Blanche's attention to it. Curt had seen it by then, too. It came closer and closer until it was parallel with us, then moved ahead of us. I told Kenny to stop; he was looking at it right then. It took a right angle and was coming down in front of us on the highway.

Blanche was screaming and shaking all over. She had alarmed Tisha, she was crying. Kenny had stopped the car by this time and I was trying to get outside. Blanche became hysterical as the craft moved across the highway to the right side of the car and came down to the ground. The entire area around it was lit up real bright. It was about one hundred yards away; it was very large, about one hundred feet and had a row of windows complete across it. Blanche was holding me and screaming; she wouldn't let me get outside. "I must go, the beings are aboard and I must meet them." The kids were frightened, but not hysterical. "Daddy, don't go, Mama can't stand it," said Shelia. Here was the chance I had been waiting for, to meet the beings that must surely be aboard that craft. Then the 'radio' was turned on. "Go. There will be another time, another place." The 'radio' was turned off. Surely, there must be goodness in them to realise how terrified my wife was and my kids being scared.

I would have been willing to have given my right arm that night to have walked aboard that craft and met those beings and communicated with them, but they told me it would be another time and so I would take my family home where they would feel safer and I look forward to the promise of those beings: 'Another time, another place.'"

6.4.4 Analysis of the encounter

In the absence of an appropriate aerospace monitoring system in the Pascagoula area on the night of October 11, 1973, we cannot say with reasonable certainty if there was a real alien presence which caused the experience. We can say, however, that the greater weight of the circumstantial evidence leads to the conclusion that Charlie and Calvin did experience an alien anomaly, and thus had an alien experience. We will now examine this evidence from its strongest and most compelling pieces to its weakest in order.

The strongest piece of circumstantial evidence by far in this case is the investigation and reaction of local law enforcement. When examining circumstantial evidence in the context of human behavior and reactions, one often considers the motives of individuals to act in a certain way. In a court of law, most commonly judges and juries will consider the motive of a witness to testify falsely. In this case, it would be irrational to conclude that the Jackson County Sheriff's Department would have any motivation to bolster or support the stories of Calvin and Charlie if it suspected they were false. If anything, the opposite is true, and local law enforcement had a strong motivation to probe the truth of Calvin and Charlie's stories and expose any falsity in them.

We see, however, that the opposite turned out to be the case. Applying forensic investigative techniques, officers in the Jackson County Sheriff's Department began with extreme skepticism of the truth of Charlie and Calvin's story and ended up believing that they had some kind of real experience. As noted from the newspaper clippings cited above, they also went out of their way to speak to the public on this issue, which they clearly viewed as necessary in order to have credibility with the community members they were sworn to protect. And here, thanks to the work of Calvin and Charlie in preserving contemporaneous clippings from the local newspaper, *Mississippi Press*, we know exactly what they told the public and when they did so.

From a standpoint of circumstantial trustworthiness, we can say that the reaction of local law enforcement, combined with the good sourcing we have for their public statements and reactions, presents a high indicia of reliability that they themselves believed something real happened to Charlie and Calvin, and that they did not believe it was a product of delusion, error, or deception. In a case like this, where we lack a robust aerospace surveillance system over our planet, this is the best we can do. But it is certainly a very strong indication that something real happened. Just like the U.S. Air Force and Navy Pilots whose work was interrupted by alien anomalies, we can say the work of the Jackson County Sheriff was similarly interrupted and that they endeavored to deal with the interruption in the same way they would deal with any other situation they might encounter.

Another circumstance of reliability here is the fact that the municipality of Pascagoula as a whole found Calvin and Charlie to be credible individuals. We see this in the fact that in 2019 their encounter was commemorated with a municipal marker. This kind of governmental response to alien anomalies is extremely rare, and perhaps even unheard of. It is also compelling when we consider that this anomaly presented itself in a small, close-knit community in the South where everybody knows everybody else's business. If these men were thought to be liars, it would have been known, and they would have been publicly humiliated. But the fact that this anomalous event continued to be viewed favorably in the minds of Pascagoula residents over 40 years after its occurrence, is another solid indication of reliability here.

Beyond that, we also have the circumstantial evidence from the men's hypnosis and examination by Dr. Harder and Hynek. First, we note that neither Charlie nor Calvin called these scientists into investigate. Second, we note that both of these scientists are extremely well-credentialled and had direct experience in the field of alien anomalies itself. Third, we have the results of the investigation itself. The men were put under hypnosis and displayed what Harder and Hynek perceived to be powerful emotional reactions which, from a standpoint of conventional medicine, would be extremely difficult to fake if they were under hypnosis. This, combined with the Sheriff Department's conclusion that the men were being honest, is yet another circumstance of reliability here. Both Hynek and Harder were convinced that something real occurred here, and although they were reluctant to say exactly "what" it is, they were willing to public vouch for the men's credibility.

Third, we have the conduct of both Calvin and Charlie themselves. The reactions of both men speak to sincerity and makes it difficult to conclude that they were lying or fabricating their reports. For Charlie, we could say that, although the experience was certainly paradigm-shattering, he adapted to it and it made his life more interesting. He was willing to speak about his experience and thought he was serving an important function. But he never acted in a "messianic" way about it, but rather more like a humble messenger. There is no evidence that Charlie was afflicted with grandiose ideation of any sort. While Charlie clearly believed his message was important, and he was willing to share it, he also maintained a normal, productive life as well. Also, we are inclined to take a man like Charlie Hickson seriously since, before his alien encounter, he was a survivor of serious, life-threatening combat situations in the Korean War. Anyone who has survived such experiences and has been able to integrate back into normal, civil society certainly deserves our respect.

When we compare Charlie's reaction to that of Calvin, the indicia of reliability in this situation becomes even clearer. It took Calvin nearly 45 years to come to terms with his alien experience in a productive and healthy way. Unlike Charlie, he was not willing to talk about his experience in public until the publication of his book and is only known to have made one public appearance, on the *Mike Douglas Show* on December 31, 1973. Other than that, he stayed relatively silent and shunned publicity until the publication of the book.

Yet, in spite of the fact that Calvin's behaviors indicated he did not want to discuss his alien experience, and perhaps wish it had never happened, he never once claimed it had been hoaxed or fabricated. Given the fact that Charlie was speaking about the alien experience in public and keeping the story alive, if Calvin wanted the story to go away, he could have very easily said it was all just a hoax and a lie to end it. But, in spite of the distress it caused him, he never did. And, ultimately, he ended up embracing it. This kind of behavior is inconsistent with an individual who is making up such a story. Combined with the totality of the other circumstantial evidence, it again bolsters the reliability of this encounter.

Lastly, there is another class of evidence which could have had great value if there had been a robust aerospace monitoring system in place at the time; the reports of other witnesses to alien anomalies at or near the time of Calvin's and Charlie's encounters. As noted above at footnote 68, there were widespread rumors of alien anomalies around the time of the reports which affected the behavior of the local populace in a very noticeable way. Beyond that, there were many other local sightings around the same time as well. A few were reported to the Jackson County's Sheriff's office at the time, but many others weren't reported until later. Again, without some system of direct, physical detection of alien anomalies, these reports are of questionable value, as they could have been the product of suggestion. But in the context of the other circumstantial evidence, they at least carry some weight.

In sum, the circumstances surrounding Pascagoula encounter carry a good weight of reliability and indicate that something real likely occurred. What exactly occurred, we have no real way of knowing due to the lack of an active, observational regime in place to capture these kinds of anomalies. Until we do, circumstantial evidence of the nature presented in this case, will be the best form of proof we can gather.

6.5 Alien experiences from the standpoint of scientific thought

As was noted above in **Section 2.12**, the term "alien anomaly" is used in two ways throughout this work: on one level, it means the anomaly is simply strange, unexplainable, "other", without declaring it to likely be a real alien. In a second sense, it is referring to an alien anomaly as an anomaly likely induced by the presence of a real alien (such as in the instance of the tic tac). Since we don't have a sufficient observational regime at present, the rule proposed for declaring an anomaly to be likely caused by a "real alien", is when the direct and circumstantial evidence, considered as a whole, make it more probable than not that the anomaly was not caused by a human, was technological in nature, and appeared to be the product of some deliberate decision-making process which we do not understand.

When alien experiences are considered from the standpoint of scientific thought, our lack of direct detection regimes should make us leery to conclusively declare any anomalies to be caused by real aliens. Even in the case of the Pascagoula Encounter, where it seems very likely that an alien anomaly was present, we should be exceedingly cautious about the categorizations we employ lest we mix the type of proof we require from scientific process with proof that is sufficient in the "non-scientific" world of everyday human beings. Thus, from a standpoint of scientific thought, we should view the alien experience expressed in the Pascagoula Encounter in the prosaic sense—i.e., as an experience that is simply "other", "strange", or "unexplained"—rather than treating it as an actual alien.

When we approach the experience from this angle, we may see it in a different light and see different ways to engage it. Instead of seeing the experience as proof of the presence of an alien, we see it as proof of the human reaction to an alien anomaly. From the scientific perspective, at least, proof of an alien anomaly would come through some form of direct detection as happened in the case of the tic tac, but not through the testimonial reports of a human subject. Such proof would be insufficient to establish scientific fact, as only more direct forms of measurement would suffice in such cases.

Instead, the human report would be viewed as a potential point of observation. It would tell us that this is a location where an alien anomaly appeared, and these are the known circumstances under which it did so. Also, in the case of the Pascagoula Encounter at least, the evidence shows that reports of alien anomalies occurred over a period of several days in and around the area. This anecdotal evidence confirms for us the utility of having a surveillance system in place as has been recommended in this work. It also shows us that the deployment of this system needs to be widespread, because we have no idea why exactly the alien anomaly occurred in Pascagoula with these two men, showing us a lack of predictability concerning the appearance of these anomalies.

We could also look at the experiences of these men as well and see that, from a scientific point of view, it would be very helpful to have some meaningful social networks in place through which these alien experiences can be discussed and socialized. Imagine, for instance, that when this anomaly occurred, there were groups in Pascagoula where people such as Calvin and Charlie could share their alien experiences in an anonymous, confidential setting. This could be similar in nature to modern-day "12 step" groups, where people from all walks of life and all socio-economic backgrounds and find help and supporting in dealing with the disease of addiction in its various forms. In these settings, people can show up at such groups at any time they choose and are allowed to maintain their confidentiality when sharing their experiences with addiction. These groups allow people facing difficult, vexing, and troubling issues, to find solace and support among other people.

Imagine if there were such groups dedicated solely to people who had alien experiences. It is possible that the experiences of Calvin and Charlie could have been more easily integrated into the community of Pascagoula. It would have also perhaps made it more likely that others who had possibly witnessed an anomaly around the same time and place, could share their experiences as well. This might allow people to compare notes, so to speak, and possibly increase collective understanding and knowledge about the specifics of the alien anomaly itself. Also, it is possible that other people who might have come into contact with the anomaly, and who had some alien experience of their own, might also be willing to share what they have experienced. This could amplify the benefit and meaning of the encounter in the community as a whole.

Aside from having a possibly therapeutic value (which is the kind of thing that any well-meaning observer would be concerned about in this type of situation), reports of the alien experiences could also be used as clues for setting up observational regimes and could give us insights on the kinds of experiences that people are reporting. Let us treat such experiences in way that alchemical researches were handled in the time before Dalton and Mendeleev. We could say that, although the experience lacks the level of scientific confirmation that we would like to see, it still tells us something about alien anomalies. A forensic consideration and recordation of such experiences could help us begin to build a catalogue of alien experiences that could be developed over time. Perhaps, then, one day when we had better systems of correlating these experiences with physical occurrences, we would be in a better position to develop real knowledge.

6.6 Alien experiences from the standpoint of metaphysical, religious, or spiritual thought

If we decide to treat alien experiences as being the product of a real alien encounter, reason and consistency requires us to approach them from the standpoint of metaphysical, religious, or spiritual thought, as opposed to a scientific one, because this is the only way in which their truth appropriately resonates. If we wanted to, we could decline to consider such experiences at all, unless and until "extraordinary proof" was presented. This is the conventional approach, and its limitations and infirmities have already been discussed.

If we are willing to provisionally accept the alien experience as being real to the human, without necessarily being real in the scientific sense of things, we are not seeking to establish its ontology, but to simply explore its phenomenology. Understanding that our acceptance of alien anomalies as true in this way is only provisional in nature, and is not something that we can validate as scientific fact, we are open to consider the human side of these experiences, regardless of whether they were actually caused by real aliens. In other words, we are accepting and exploring the human truth, without accepting it as objectively true.

Charlie Hickson's alien experience created a sense within him that he was a messenger of some sort, between aliens and humans. While his alien experience opened his mind to new possibilities, it did not prevent him from also having a normal and productive life, even though he did claim to be visited by aliens on subsequent occasions and even receive telepathic communications from them. For Charlie, it seemed he possessed the innate capacity to integrate the strangeness of his experience with his normal life; indeed, the integration was so well accomplished that he was ready and willing to make a conscious decision to enter into an alien spacecraft. Any fear he might have experienced was replaced with a strong and resolute curiosity. Charlie appeared to be able to competently straddle and reconcile the inconsistencies between the ordinary human world and the alien one in which he had been thrust. He appears to have been able to keep these worlds separate in his mind and human experience. Hickson's approach stands out as an excellent paradigm for dealing with alien anomalies and alien experiences.

Whitley Streiber, by contrast, did have his reality altered in a very radical way, and had all the lines of normalcy blurred for him. The following citation, taken from *Communion*, is a paradigmatic example of the kind of self-examination and questioning that Streiber's alien experiences induced within him.

> What might be hidden in the dark part of my mind? I thought then that I was dancing on the thinnest edge of my soul. Below me were vast spaces totally unknown. Not psychiatry, not religion, not biology could penetrate that depth. None of them had any real idea of what lives within. They only knew what little it had chosen to reveal of itself.
>
> Were human beings what we seemed to be, or did we have another purpose in another world? Perhaps our life here on Earth was a mere drift of shadow, incidental to our real truth. Maybe this was quite literally a stage, and we were blind actors.

> To gain some semblance of control over myself, I decided to make an inventory of possibilities. I sat down at my desk and began to write.
>
> Even if the visitors were real, there was no reason to believe that they were simply creatures from another planet.
>
> I speculated it could be that the "visitors" were really from here. Certainly the long tradition of fairy lore suggested that something had been with us for far more than the forty or fifty years since the phenomenon took on its present appearance. The only trouble with this theory was that what has been happening since the mid-forties seemed more than just a little different from the fairy lore. Now there were brain probes and flying discs involved, abductions and gray creatures with staring eyes. Surely no change had taken place in the human psyche extreme enough to account for such a radical change in the appearance of the fairy. And yet there was undoubtedly something here . . . I thought perhaps the visitors were somehow trying to hide themselves in our folklore.
>
> Another thought was that the visitors might really be our own dead. Maybe we were a larval form, and the adults of our species were as incomprehensible to us, as totally unimaginable, as the butterfly must be to the caterpillar. Perhaps the dead had been having their own technological revolution, and were learning to break through the limits of their bourne.
>
> Or perhaps something very real had emerged from our own unconscious mind, taking actual, physical form and coming forth to haunt us.

There are many more examples of these kinds of internal reflections in Streiber's work. Streiber's mode of recounting his experience is similar to a class of experience that has been going on for some time in the modern human consciousness, albeit in a different context. Consider, for instance, this citation of a so-called "religious experience" from William James' work *Varieties of Religious Experience* (1902) in a chapter entitled, *"The Reality of the Unseen"*, where Prof. James provides an account of an experience which an intimate friend of his supposedly experienced:

> I have several times with the past few years felt the so-called "consciousness of a presence". The experiences which I have in mind are clearly distinguishable from another kind of experience which I have had very frequently, and which I fancy many persons would also call the "consciousness of a presence". But the difference for me between the two sets of experience is as great as the difference between feeling a slight warmth originating I know not where, and standing in the midst of a conflagration with all the ordinary senses alert.
>
> It was about September, 1884, when I had the first experience. On the previous night I had had, after getting into bed at my rooms in College, a vivid tactile hallucination of being grasped by the arm, which made me get up and search the room for an intruder; but the sense of presence properly so called came on the next night. After I had got into bed and blown out the candle, I lay awake awhile thinking on the previous night's experience, when suddenly I FELT something come into the room and stay close to my bed. It remained only a minute or two. I did not recognize it by any ordinary sense and yet there was a horribly unpleasant "sensation" connected with it. It stirred something more at the roots of my being than any ordinary perception. The feeling had something of the quality of a very large tearing vital pain spreading chiefly over the chest, but within the organism—and yet the feeling was not PAIN so much as ABHORRENCE. At all events, something was present with me, and I knew its presence far more surely than I have ever known the presence of any fleshly living creature. I was conscious of its departure as of its coming: an almost instantaneous swift going through the door, and the "horrible sensation" disappeared.

This type of account, coming from 1884, is well before the dawn of the alien paradigms in which someone like Streiber can process his experiences. But if one takes time to read any of Streiber's work, one will see that many of the experiences that Streiber has as a result of what he calls "The Visitors", are eerily similar to what James' friend experiences in the late nineteenth century.

Streiber's experiences can be placed within a larger catalogue of human experience which predate the emergence of alien paradigms and simply speak to the human capacity to have one foot in the normal world, and another in a world that is alien. What is different in the context of Streiber's experiences, however, is the possibility that all of this weirdness is not caused by "God", "Spirit", or some universal "Creator", but rather by beings that may stand in relation to us in a similar way that humans stand in relation to, say, modern, domesticated canines. We and our dogs do share a world together, but there are so many aspects of our world that they could never understand. This is a new experience for our species—to learn what it is like to be in the presence of a higher order being that is in the world with us, and not in the supernatural realm. *See* **Section 2.6** ("the 3 orders"). Such knowledge differs from the more normal "spiritual" or "religious" forms of revelation, insofar as we attribute it to a discrete being, as opposed to the Author of Existence itself, herself, or himself.

Charlie's alien experience is far different, and maybe more appropriate to a scientific perspective. As we have discussed previously, one of the goals of science is to take objects from the invisible orders and integrate them into the normal world that changes the normal world, while still allowing it to be conventionally relatable. Charlie's approach to alien anomalies may be far more radical than Streiber's, precisely because it is less weird, and more real—so real, in fact, that Charlie was ready and willing to come face to face with alien beings in the conscious daylight of the real human world. Such an encounter should be the lawful goal of a genuine scientific endeavor due to its power and ability to transform conventional relations and create new, alien forms of normality. Charlie Hickson stands out as a true pioneer in this arena. Charlie Hickson was no terrestrial redneck, that's for sure.

6.6.1 Why aliens should not be placed in the supernatural order

When dealing with alien experiences, people often times speak and think of them in a way that suggests they fit naturally into the supernatural order discussed in **Section 2.7.** It is understandable why this would be the initial reaction, but it is also creates an unnecessary form of cognitive distortion. Within the conventional framework itself, this placement is not a good fit because aliens, even if they are beings from a higher order who display magical powers, are simply not the same as the beings we naturally place in the supernatural order itself.

As was noted in **Section 2.5**, the existence of the supernatural order often times seems necessary as a conceptual construct to explain the conventional features of the world. Even many great scientists, such as Einstein and others, feel the need to posit this order to help them make sense of the world. They do not posit this order because they are foolish or weak minded, but rather because they feel incapable of expressing a complete operation of the world without it.

One key feature of the supernatural order is its tenuous and sporadic relationship to the conventional one. By definition, supernatural beings are not a regular and normal aspect of the conventional world. It is precisely because they are unconventional and sporadic that they form the supernatural order in the first place and are imbibed with powers and attributes that help explain how this world is held together. While we can see why aliens would most naturally appear to the human mind in this way, deliberative analysis indicates that their placement there is not a good fit because, unlike the supernatural order, the alien order is in some sense contiguous with the normal world (in this connection, increasing our knowledge of measurement and "hard data" is one of the ways these anomalies are brought into the more normal world). While we might say that aliens are from another time in the sense that they are the product of an advanced consciousness that has evolved over spans of time that exceed our own by several orders of magnitude, they are nonetheless in our type of embodied time along with us. As far as conventional understanding goes, they too are subject in some sense to the laws of life, death, and time.[75]

As was noted at the outset of this work, the notion that there is an "unseen creator" behind the phenomena of the world is quite logical due to the numerous gaps in our understanding of the discoveries that we make. As we continue on this path, we keep finding more and more complex structures in the world that appear to be "created" and which we know we didn't create. The way in which we experience our world strongly suggests that the world of hidden and invisible structures is very ancient and came into being well before we humans were even in a twinkle in the eyes of the

---

[75] The case of interdimensional beings stretches this statement to the breaking point because such beings are presumed to exist in a manner that defies normal temporal parameters and the conventional rules of embodiment itself.

cosmos. The human mind is therefore naturally drawn to the idea of a creator based solely upon logic and experience with reference to the idea of creation in the conventional world itself.

Starting from this initial impression of things, as well as their own personal experience, humans have created an immensely divergent range of views as to what this creator, or creators, must be like. Clearly, all of these ideas about this creator are human generated, even if the motive or inspiration behind the generation was divine. Such systems are the real products of human experience and thought, existing in the world of temporal imperfection. They represent lower forms of knowledge in relation to the fully formed higher orders that created objects such as atoms, DNA, subatomic particles, galaxies, and flagellar motors in the first place. Our ability to discover and detail these objects, while impressive, is still inferior to the unseen order that created them.

While some claim to have visibility and understanding into this unseen or supernatural order, such knowledge clearly lacks the regularity that is expressed in more conventional forms of knowledge, such as mathematics, chemistry, optics, and the like. The origination of the higher order regularity, and how it comes about, remains a mystery to us (just think about the real origins of hyperbolic space, of which we know nothing). Even though we can sense our inferiority in the face of these facts, we can also see within ourselves elements of higher orders of knowledge which we would assume any real creator would have. At the present time on earth, for instance, we see no other creatures that display intellectual, productive, and creative traits in the ways that we do.

As argued previously, however, in **Section 3.3.3.2.1.4,** we do not have an *a priori* basis upon which to assume that we are the only terrestrial creatures who could attain to such traits and capabilities; it is simply the case that we are the only ones at present on earth who have done so yet. In a 100 million years, the case could be different, and other species which today lack the same higher order capabilities we do (such as certain species of birds or insects) could possess them. Nimoy's Dictum says, at least, that we cannot rule this out. In any event, we see on earth at present no peers with respect to our intellectual, creative, and productive capabilities, and thus rightfully view ourselves as beings of the "highest order" here on earth. We then naturally infer a similar relation exists between us and the creator(s).

Our own experience on earth therefore provides a firm basis for us to realistically believe in different orders of beings, with some being higher and others being lower. Given our position on earth when we became aware of our position in the terrestrial system of life, we can see why many would naturally and initially assume that we were created in the image of "God". At some point, this viewpoint ripened further into an idea that we could even understand God. But in a practical sense we can see today that the relationship between us and the creator(s) is very rough and incomplete as science reveals to us a barely comprehensible universe, with alien lifeforms being the newest example of such a thing. Yes, it is true that we continue to accumulate knowledge about the mechanics of our "local conditions", but the further out we go, the greater the mystery deepens, leaving us with a sense of general awe and incomprehensibility as to what the creator(s) is/are actually capable of.

It is within this context that alien experiences arise. Because aliens seem to possess powers far beyond our own and emanate from regions of the universe of which we know nothing concrete, it is natural that their initial presentation to the human mind would partake of certain elements of godlike divinity. As a result, many people stay stuck at this level of analysis and relate to them solely in this way. Here we are suggesting a different approach in the interest of relating to aliens in a mature and realistic way. This approach acknowledges their initial placement within the conventionally determined "supernatural" order but suggests that we should ultimately place them within orders that are more consistent with conventional knowledge about aliens itself. Thus, we should reject the idea that they are "gods" and decline to place them in pre-existent terrestrially created categories that arise from our ancient history. Aliens are arising in a new category that is wholly alien to us and which we are barely beginning to understand.

However, we still need to account at some level for the fact that these alien beings appear to have powers that are far beyond anything we can muster. In this connection, we could consider relating to aliens, not as gods or divine beings, but rather simply as "higher powers" that are potentially comprehensible to us. That is, aliens could be viewed as beings that arise within an order of life that is more temporally relatable to us than that of the supernatural order itself—even if they possess powers that allow them to move through time and space in ways that appear magical at present. *See* **Section 2.3** ("cosmic solemnity"). We see an expression of this approach in the work of William James in *Varieties of Religious Experience*, in Lecture Two of his work which reads as follows,

> There must be something solemn, serious, and tender about any attitude which we denominate religious. If glad, it must not grin or snicker. If sad, it must not scream or curse. It is precisely as being SOLEMN experiences that I wish to interest you in religious experiences. So I propose—arbitrarily again, if you please—to narrow our definition once more by saying that the word divine as employed therein shall mean for us not merely the primal and enveloping as real, for that meaning, if taken without restriction, might prove to be too broad. The divine shall mean for us only such a primal reality as the individual feels impelled to respond to solemnly and gravely, and neither by a curse nor a jest.
>
> But solemnity and gravity, and all such emotional attributes, admit of various shades; and do what we will with our defining, the truth must at last be confronted that we are dealing with a field of experience where there is not a single conception that can be sharply drawn. The pretension under such conditions, to be rigorously "scientific" or "exact" in our terms would only stamp us as lacking in understanding of our task. Things are more or less divine, states of mind are more or less religious, reactions are more or less total, but the boundaries are always misty, and it is everywhere a question of amount and degree.

Unlike the relationship to the supernatural order, which is assumed to be intrinsically inaccessible and sporadically available to us, this alien order should be relatable in a new and different way. The alien order should not be presumed to be "out of time" in the same way that the supernatural order would be, even if these alien beings appear to display supernatural powers right now. *See* **Section 4.8** ("Clarke's dictum and the scientific conversion of alien anomalies"). But still, there should be something awe-inspiring and solemn for us about the opportunity to simply stand in the presence of these mysterious, alien circumstances. And we don't have to lose our minds or go crazy in order to appreciate it, as Charlie Hickson has shown us.

Drawing again on the Copernican revision, if a person from the pre-Copernican era was transported to our present era, we would certainly move and do things in ways that they couldn't possibly understand and would presumably appear alien to them in many ways. In this scenario, the dissonance in perception and understanding that would create such an alien distance is the expanse of time, for we ourselves would know that we are not actually gods in relation to these people.

For any real alien which has made it to earth from some far away planet, we must assume that they have a far different relationship to time than we do, allowing them to transverse impossibly large distances within spans of time that are discrete within their own lifetimes, which would not consume, in other words, the entirety of their existence. So in a way, we could view the alien relationship as another way that the cosmos is showing us the depth of its time as expressed through the consciousness of other beings—this may induce in us a sense of cosmic solemnity, as opposed to numinous overexcitement. These are new types of feelings for us, new horizons of human experience. We are just now becoming capable of apprehending alien lifeforms in this kind of way and do not need to rely on the constructs arising from the supernatural order to relate to them as such.

These types of thoughts can replace our need to project supernatural qualities on alien beings, and instead contextualize them as best we can within their alien orders. We are free to think in this way when we engage with alien experiences as expressing something real in the human world. The alien experiences of people like Streiber and Hickson give us a window into this type of relationship, but they do not foreclose or direct how any of us must think of it, nor can we say in any way that their experiences exhaust what is possible in the world of alien experiences. The beings that Streiber and Hickson encountered could just be two types among an innumerable set. There could be aliens *en route* to earth as we speak, as yet to appear, which could be vastly different than these two types (which, incidentally, are just two types among the range of alien experiences that have been expressed by many different people).

6.7 *The possibilities of alien experience*

We can distinguish here different possibilities for alien experience/communication. The alien resides in a different sort of embodiment within our realm. It does have a "godlike" intelligence, but is not really a god any more than humans are actual gods in relation to their canine dependents. Aliens represent an intelligence that has been developed over a far greater, perhaps incomprehensible, time. But this is different than the gods that are "out of time", or "eternal", to us. These alien beings are in some kind of shared time with us which is not coextensive with the supernatural order as we conventionally conceive of it. By communicating or interacting with them, we can learn more about this shared time and how our experience with it might change us in unanticipated ways.

But if our platform for relation remains stuck at the level of the supernatural order, we will likely always be yearning for the same type of sporadic and mysterious connection that our ancestors have been chasing with this order since the beginning of recorded time. We must be aware of the fact that *before* we measure, observe, and draw conclusions, our thoughts and perceptions are formed by certain paradigms which color and shape those perceptions and thoughts. We must bring a different type of awareness to this level as a new *a priori* outlook that is the result of conscious deliberation, instead of subconscious impulse and accident. So why continue to pursue an ancient pathway that seems to always lead to dead ends, when there could be something entirely new which we have never even conceived of just over the horizon?

We can see these alien experiences as containing elements of the divine in them, but not in the same way that we might feel about divine interaction with a "creator". As noted above, William James defined divinity as "such a primal reality as the individual feels impelled to respond to solemnly and gravely, and neither by curse or jest". We can think of the "human" to "alien" contact in that divine way, without turning the alien into a god. We focus on the system that is created, and our participation in it. We are the "lower" intelligence in the relationship and are aware of it as such. Thus, it is our proper role to be as aware as possible of the "lowness" of our place in the order. This requires us to surrender the idea of complete understanding and accept that we cannot understand alien agendas (just like our dogs could never fully understand human agendas). And yet, by coming into contact with the alien intelligence and being aware of the overall relationship, we are certainly doing something; we are moving ourselves in a different direction.

Think of it like this. Imagine how startling it would be if the dog began to suspect that his or her human master was operating in alien agendas which he or she could never understand. The dog might be overpowered by a sense of awe, wonder, and amazement. Or imagine the lab rat that looked up from the maze one day, peering directly into the eyes of the scientist and seeing some type of awe-inspiring power residing there. This kind of feeling would express a new intelligence, even if none of the details of what it means were fully formed. The bare awareness that there exists a real order of which the dog or the rat was unaware could be a shift of enormous magnitude. A seed of a new type of awareness. Isn't this the kind of thing we should be seeking to cultivate through alien experiences instead of getting lost in the reduplication of forms of religious relationship that were formed in the infancy of our species? Why not take a chance and try something new here? After all, the world of alien experiences should be new terrain, and not simply an opportunity to reproject old, terrestrially based ideas in shiny new packaging.

6.8 Conclusion

Humans of the 21st century face a far different, yet similar, dilemma to those in Copernicus' times which we have not yet figured out. An exploration of alien experiences *qua* alien experiences is just one more vehicle we can use in this exploration, even while acknowledging that these experiences lack the type of normality and stability that we expect and demand from conventional scientific operations and facts. In Copernicus' day, that dilemma concerned how one could actually accept that the earth moved in space when the "known facts" indicated this was impossible. We today can see the interpretive structures that color these facts in a way that those people could not. Seeing the real interpretive structures in the strange facts of alien anomalies, however, is something that we struggle with today.

Today's human reality presents a world where aliens are objects of the imagination and fantasy in the conceptual realm yet sometimes appear to present "real effects" in the embodied world in which we live; and yet, establishing the reality of these alien anomalies at a level of basic conventional reality has still not yet occurred (just as people in Copernicus' time were unable to accept the basic reality of the earth's movement around the sun in space for quite some time). Alien experiences therefore naturally incline us to project ideas from the supernatural order upon them because the filters in our minds, our ancient paradigms, predispose us to view things in this way. But the record of human achievement teaches us that we need not be bound by the paradigms of the past. There are other ways, new ways, in which we could approach alien experiences, instead of simply dredging up analogies from the past and reprojecting them.

On experiencing an alien anomaly, we can reside in a basic sense of solemnity that one is witnessing the work of an alien being in the human world. To reside in this way, we need not speculate upon, or anthropomorphize, the anomaly, but simply and basically accept that one has just witnessed the intrusion of an alien intelligence into the human world, watching with a sense of cosmic solemnity. One can marvel at the fact that the universe is amazingly old, amazingly large and contains beings we have not even imagined. Experiencing an alien anomaly is an invitation to explore those possibilities as they intrude upon our little human world. One is, of course, free to reject the invitation if one chooses.

7.0 Conclusion

The core argument of this text has been the proposition that we today stand in a similar relationship with respect to alien anomalies that the people of Copernicus' time stood in relation to the "true" nature of the earth's position in space. To the people of the 16th and 17th centuries, the idea that the earth moved around the sun in space was an absurd, unhinged idea, seemingly contradicting everything they knew about the world as well as their direct, lived experience. People of that time had been conditioned to believe that the stability of the world and the facts of life were dependent upon the earth's central and stationary position in the universe (both physically and religiously). Removing this postulate without presenting a valid substitution to explain how the world worked induced a neurotic counter-reaction and mobilized fierce cognitive resistance. It was not until people could look through Galileo's telescope at the movements of the moons of Jupiter that they could begin to accept Copernican truth and move on from there.

For reasons that appear superficially sane, conventional systems of thought and perspective have great difficulty accepting that intelligent aliens from other planets, solar systems, and/or galaxies could be actively visiting earth right now—after all, we have been conditioned by science fiction fantasy to believe that the only time an alien and a human would actually interact is in the context of science fiction fantasy itself. The reason for this belief, however, has nothing to do with real knowledge, and is primarily an act of subconscious projection. We have adopted a belief, without critical reflection, which says that all alien beings that can practice the arts of science and manufacture are part of a universal scientific community of which we are members, and that all such aliens, if they came to earth, would want to be our friends and spirit guides. We want to believe that such alien beings would be ready to become involved in our troubled world and rescue us from all of the problems we have created for ourselves. These ideas and beliefs, however, simply interfere with the development of genuine knowledge about alien anomalies. There is no real basis in fact for these positions whatsoever, as the cosmos is far deeper and much stranger than our human imaginations can conceive.

Because the idea of alien visitation is presumptively so unbelievable, conventional understanding has decided that if aliens were really here, the signs of the alien visitation would be global, irrevocable, and obvious, requiring no "extraordinary" investigations, and would appear in ways that are consistent with proof from a science fiction movie (i.e., Sagan's "logbook of an alien captain"). However, these viewpoints are not the result of rigorous analysis themselves, and instead are simply subconscious projections in response to the potential presence of a numinous object. They are not reliable guideposts upon which to base our judgments.

In truth, we lack the knowledge base to make any reliable declarations about the intimate and personal characteristics of unknown beings who have evolved on planets we know nothing about. Thus, we have no basis to conclude that if aliens arrived on earth, they would behave in any particular way or have any great interest in making their presence known to us on a species wide basis (or that they would even understand what that might mean to us, or how they might do it). Just as we don't really know how to directly communicate to bees in ways that members of the hive would directly understand, it is possible that aliens might have just as much a hard time relating to and understanding us, as we might have relating to and understanding them.[76]

Furthermore, the narrative that aliens would arrive at earth and "disclose" their presence at once to humanity appears to be a projection of the things that we would like to see as a supernatural fix to the myriad of unresolved and seemingly intractable problems we face as a species, but not the product of a rigorous, sober, and disinterested analysis. An exacting consideration here shows that we have no good reasons to believe that aliens would behave in any particular way that would make sense to us or that would fall within the realm of our terrestrially conditioned expectations. Thus, we should keep an open mind on the topic and study the anomalies that present themselves in the greatest detail possible instead of

---

[76] Within the terrestrial order itself, there are innumerable examples of alien orders interacting. Think of the human relationship to insects, reptiles, and birds. While there is some dimensional overlap between them, the internal workings of these creatures are alien to us, just as we are alien to them. Thus, one could plausibly state that "alien" relationships are actually standard features, and not simply bugs, within the order of life itself as we know it.

fantasizing about them, or simply ignoring them and hoping they will go scurry away back into deep space and leave us alone.[77]

While conventional resistance to admitting live, active aliens into the field of direct human experience may seem eminently rational, it is only rational to the human mind because it is consistent with the default experience of human reality thus far, which shows that humans share the earth with other humans and animals, but not with other aliens, and that aliens are only real in science fiction fantasy. But if it turns out to be the case that aliens really are causing at least some of the alien anomalies that we have perceived, this rational skepticism will only be rational in the same type of way that common sense resistance to the teachings of Copernicus was rational before the direct experience provided by Galileo's telescope. It is true that in a world where there was no such thing as Newtonian or Einsteinian physics, and the systems of reason had been built upon Aristotelian notions of space, time, and matter, resistance to the ideas of Copernicus was entirely rational, and acceptance of Copernicus' ideas seemed insane. However, this feeling of rationality, while compelling, does not seem very accurate to us now. There was a deeper, Copernican truth that was lurking beneath the surface that would not be dissolved by collective ignorance.

Considering the example of the Copernican Revision in relation to alien anomalies, perhaps one can say that any sufficiently revolutionary paradigm shift will initially appear insane to those who have been conditioned to believe that their present paradigms are complete, rational, and necessary. But these feelings are not a function of objective truth, so much as human attachment to a particular paradigm and a reluctance to let it go for something new. The trick is to see beyond these false feelings and allow the deeper truth to emerge. This, of course, is a complicated enterprise,

---

[77] We understand that these positions are very difficult to digest. This is what we went to such lengths to analyze the epistemological limitation of the safe distance paradigm in **Section 3.3.3**, as well as **Sections 5.3** through **5.5**, and other places in this work. If one considers this analysis in light of the Copernican Revision (first discussed in **Section 1.3**), one can see how difficult it can be to get at these deep collective biases sometimes. Again, people of the 16th century like Jean Bodin weren't just "stupid" when they said that Copernicus' ideas did not make sense; nor are people today just being "stupid" when they refuse to acknowledge and explore alien anomalies—they are, however, being neurotic.

especially in cases where the objects of thought are not confined to the laboratory setting and are part of the fabric of reality itself. In those cases, the declarations of the specialized communities or knowledge bearers (e.g., scientists) are insufficient to move the needle due to the cognitive and perceptual inertia created by the conditioning effects of the dominant paradigm in operation at the time. It is not enough just to give observers information which is more "correct" due to the manner in which paradigms are embedded deep within the structure of cognition and perception itself through repeated and sustained experience. In those cases, this inertia can only be countermanded by the possibilities of direct experience itself (i.e., looking through Galileo's telescope, or through other spectacles of demonstration such as the Magdeburg hemispheres).

7.1 Working out an alien paradigm

We can see why the possibility of aliens surreptitiously lurking around earth would be an insane idea which appears completely contradicted by the rational perceptions of daily life. But this is just our human perception of things, it is not an irrefutable or objective presentation of truth. If a person believes very strongly that aliens are not present in earth's aerospace, then any indications of their presence will be "hard to believe" by definition and readily dismissed (this would be especially true of academic scientists who, unlike military pilots, spend very little time in the skies, and are free to ignore alien anomalies if they want to).

Additionally, if aliens are utilizing technology that appears to defy our normal laws of gravity, friction, and resistance, it will naturally be difficult for us to perceive them in the first place because we are not conditioned to perceive objects which move in that way just as denizens of the Copernican era could not understand how birds would fly, buildings would stand, canons would fire, and water would successfully pour into flagons, if the earth was really spinning around in space at tremendous velocities. In a certain sense, one could say that the paradigms of motion with which we are familiar, and which some of us think are universal in any cases involving motion, have conditioned us to perceive in certain ways and that objects which move inconsistently with these paradigms will be very difficult for us to perceive at first. Just as the Aristotelian could not see the same pendulum that the Galilean saw, we might have great difficulty in perceiving these alien anomalies in a clear, unobscured light due to the strange and incomprehensible nature of their movements. Thus, it may take extra time and special attention to retrain our perceptions and cognitions to be open to these alien forms of motion, as these anomalies may challenge the direct experience of physicality itself.[78]

As a first step in retraining our cognitions, we could refrain from gross speculation about what aliens might be like and might be doing, and stay focused on studying the alien anomalies that are going on around us, drawing rational inferences as we are able. Thus, we should consider taking a page from the evolution of clinical medicine in the late 17th and early 18th centuries when clinicians put considerable thought and effort into retraining and recalibrating their perceptions as an integral part of their course of study, and abandoned the fixed ideas of nosology which prejudiced their perceptions in advance. (here, we consider inquiries into alien deep space megastructures, Dyson spheres, Fermi's paradox, Sagan's Dictum, and the Drake Equation as the analogy of an unproductive nosology with respect to the study of alien anomalies).

In this connection, return to the example of the tic tac yet again. Instead of wondering about what is "inside" the tic tacs, let us focus instead on the observable facts and circumstances, and the rational inferences we might draw from them in light of the conventional categories of knowledge. There is but one of three explanations for this anomaly: 1) natural forces; 2) human activity; or 3) aliens. Of the three, which is most plausible given what we know? Certainly, natural causes are highly implausible because it would distort and violate the meaning of the term "natural" itself to speak or think in this way, resulting in pure absurdity and nonsense. As we noted at the outset of this text, we moderns draw sharp distinctions between natural forces, human forces, and supernatural forces. This is the language that we use and how we think about the world, regardless of whether our thinking is "right" or "wrong". Natural forces are depersonalized and move at a rhythm and pacing that is specific to their nature. We speak of

---

[78] A precise application of the Copernican Revision shows that the most bedrock principles of reality itself could be called into question. The notion that our idea of physicality itself can be challenged is extremely absurd, and perhaps unsettling; and yet, this is exactly the kind of thing the Copernican Revision itself challenged in the minds of the people of the 16th century.

cosmological time, geological time, the human heartbeat, seasons, the timeframes of the chemical or subatomic worlds, the timing of animal instinct, and so on. There are many different manners in which a wide range of phenomena propagate, we might say, through the "medium" of time; the force that seems to drive these types of phenomena does not emanate from the same type of force that is responsible for human choice and decision, which seems to arise for us from a different level altogether. As noted in **Section 2.0,** we draw a sharp distinction in our minds between the phenomena of natural forces and the phenomena which are generated by our fellow humans (and sometimes animals) which appears to come through a timing process directed by forces such as intention, will, deliberation, and choice (i.e., the products of a "living self"). We base this distinction on the experience that is most personal and immediate to us, and which is the driving force behind all of forays into knowledge: the living presence of an active self.

The movement of the tic tac is fundamentally inconsistent with the motion of any natural phenomena as we ourselves would understand and use that term: the sharp turns, the sudden disappearance as it shoots off into the distance, the sense that it was responding to the movements of the pilots, the sudden appearance at the hold point. The nature of these motions naturally suggests to the mind a force which is not impersonal and natural, and which adapts to the changing conditions of time in a rapid, intentional manner. The timing and presentation of these motions are fundamentally inconsistent with the category natural as it is used in conventional discourse. Instead, they are similar to the activities of an active, living self, such as we ourselves. If an observer wishes to categorize the tic tac as the result of a natural cause, they are essentially destroying the idea of a natural cause altogether and turning the category "natural" into the ancient category of "supernatural", imbibing natural forces with the powers of supernatural abilities. No rational observer should even consider such an absurd result.

What about the category of human cause? There are two possibilities here. The first is that the tic tac was the product of error or delusion and that the witnesses essentially hallucinated the entire event. Setting aside the fact that the navy pilots are the most highly trained aerospace observers in human history, would have no reason to fabricate their sighting (after all, at the time they were getting ready to go to war and were not just screwing around), and the fact that multiple observers from multiple vantage points all corroborated the sightings, the presence of the FLIR image of the tic tac demonstrates that the tic tac was indeed a real object. The data gathered from the FLIR camera came from the non-optical band of infrared radiation; that only happens if a real body is emanating actual radiation. No rational observer would suggest that a human mind can project a delusion into physical space which would then emit infrared radiation mimicking the type of radiation that one would see from a real body. The suggestion would be absurd, insane, and something one would expect to see in a Marvel movie, not in real life.

The second possibility is that the tic tac was really the result of some secret research program developed by human beings. But is this plausible? Recall that the tic tac was seen over 20 years ago, and that since that time a sighting of this magnitude has not been publicly repeated. First off, if this was the product of humans, which group would have the resources and capability to develop it? Among human militaries, the only three who would have a chance at doing so would be America (which includes the European allies), China, and Russia. If the tic tac was the product of the American military, then why on earth would the American military go to such lengths to develop a secret program, not tell the U.S. Navy about it, and then have it inadvertently exposed to the world in a way that reveals the technology without then ever using it, gaining no strategic military advantage? This makes no sense. The same would be true with respect to Russia and China, especially given the rise of global tensions in the early 21st century. If any one government had a technology of this nature it would be what U.S Representative Tim Burchett called "a game changer", and there is no doubt they would use it in some way to gain military superiority. After all, why develop this kind of thing if not to one day use it, and what better time to use it than now when more and more nations are engaging in open warfare with each other? It would be patently absurd, and contrary to the intentions and behaviors of military thinkers, to suggest that the tic tac was the result of a secret military program.

Similarly, the idea that the tic tac was developed by a non-military group of humans is also absurd and implausible. The amount of money and time it would take to develop such a technology would be astounding and would no doubt cost billions and billions of dollars spent over a long period of time. Very few people on earth had those resources in the early 2000's and would have had to have had them for decades before that in order to develop this technology. And what scientists would they have even enrolled to develop such technology in the first place? Today's brightest scientists cannot even explain how the tic tac is doing what it does, and many of them find it so hard to believe from a standpoint of physics, that they doubt it was a real occurrence and have no interest in studying it or even considering it.

While it might be "less extraordinary" to say that the tic tac was the result of natural, human, or "unknown" or "unidentified" causes, basic logic, reason, and the *categories of conventional thought themselves*, show that the "alien hypothesis" for the tic tac is far and away the most plausible choice here. Again, Cocconi and Morrison's basic premise—that exoplanetary science, evolutionary theory, and human technological achievement strongly imply the existence of intelligent aliens on other planets—remains sound and today has the status of an article of scientific faith. Thus, "real aliens" have been added to the legitimate categories of possible scientific causes for some of the phenomena that humans might see. This doesn't mean that by saying the most plausible cause of the tic tac anomaly is an "alien" we have "proven" aliens are "real"—all it means is that from a menu of options available to human cognition, the most exotic one just happens to be the most rational one as well. The observations of the navy pilots, along with the direct evidence that was gathered, strongly suggests that the tic tac was powered by a technology which is not limited by the conventional transfer of momentum theory of propulsion and is not restricted to the conventional restraints of friction and resistance. The anomaly appears technological in nature, but the technology does not appear to be of any form of technology that a 21st century human would know how to make or use. The tic tac's movement is so unbelievable that it seems like magic, requiring us to invoke Clarke's dictum that any sufficiently advanced technology would, by definition, appear to us in this way.

The paradigm that accepts this line of logic would be, formally speaking, an "alien paradigm". All this means is that instead of denying the possibility of alien presence as the true cause of the phenomenon, or always saying "unidentified" or "unknown", the scientifically inclined thinker attempts to construct a positive network of meaning around an idea which seems plausible, even if poorly understood (or not understood at all). This is a non-arbitrary basis upon which a reasonable "article of scientific faith" can be founded; thus, proceeding under this faith has a real chance at leading to some forms of real knowledge. Kuhn teaches us that this is how science actually moves, or "converts", from the pre-science phase into normal scientific activity. This is very different from the current approach to alien anomalies which is either mired in the quicksand of pure negation (where observers will only say the anomaly is "unknown" or "unidentified" and never attempt a positive description), or stuck in the SETI paradigm's arbitrary sets of unexamined assumptions (that only lead to neurotic, unproductive forms of fantasy speculation).

7.2 *A vision of disclosure*

If we compare the engagement with alien anomalies to the Copernican Revolution, we can see that we are right on track for the kind of shift into a more positive phase of science at this point in human history and that the neurotic indecision that has plagued conventional thought with respect to alien anomalies checks out. While some may lament the fact that we did not follow the Project Sign approach in 1947 and instead embarked a course of suppression, it is entirely understandable that rigid military thinkers would recoil at the possibility of acknowledging alien presence—and, more importantly, we can change the way we approach these anomalies if we really want to today. After all, the embrace of real, live, active aliens by human beings would be far more monumental and paradigm-shattering than the Copernican Revolution was. It could change the course of human development and understanding in ways that are just as unimaginable to us today, as the rise of Newtonian, Einsteinian, and quantum physics would have been to someone in the 16th or 17th century. Imagine how radically different the human world would be if aliens one day became a feature—not a bug—of conventional human existence, and if alien anomalies were accepted as a basic fact of life for our species, occasional bearers of knowledge from regions of the cosmos that are far beyond our terrestrial understanding. With each acknowledgement of these alien anomalies, we learn more about the depth and strangeness of the universe itself, integrating this bizarre understanding incrementally into the human world as we know it.

We can replace the current obsession with revelatory disclosure with a rational "vision of disclosure" that pursues this avenue of knowledge in a realistic way. There could be three main aspects to this vision. The first, is scientific. Essentially, this would require the study of alien anomalies in the same way that other scientific objects are studied. It would require humans to treat the seas and skies as the laboratory and invest sufficient time and resources in a wide-scale survey of anomalies, using the latest technologies. It would also require a completely transparent and publicly integrated approach. Presently, those scientists who are keeping an open mind on alien anomalies, and are actively studying them, believe that peer reviewed papers and declarations from scientific specialists will be enough to move the needle on alien anomalies. The example of the Copernican Revolution shows that this view is likely incorrect. A modern-day analog of Galileo's telescope may be needed for people to decide whether alien anomalies are real.

We can easily conceive of examples of this kind of thing and how they might have a powerful effect on the collective understanding. Consider again the imaginary scientific version of the *USS Princeton* as it monitors the tic tacs off the California coast in November of 2004. As it approaches the tic tacs, not only is it not in a hurry, with an open attitude of curiosity, but it is also taking measurements of physical disruptions with advanced radar systems. While this is happening, the visual encounters, as well as the read outs from the systems that register physical activity, are being broadcast live to anyone who wishes to watch the events unfold via the world wide web. One could only imagine how tens of millions of humans, watching such an event unfold in real time and knowing that, whatever it is it is physically real, would have a stunning impact on the willingness and desire to know more. This would be a modern-day analogue of Geuricke's Magdeburg Hemispheres demonstration, and have a profound breakthrough effect on the collective consciousness of our species.

Consider a less dramatic example. Imagine that in towns across America there are towers of observation using the same technologies that were aboard the *USS Princeton*. These towers monitor the aerospace within a specific radius. Let us assume that observer A sees some sort of anomaly and either records video of the anomaly, or simply makes a voice memo of what was observed. Let us further assume that observer A is able to send geolocation data from her phone to the tower, which can then check its own data set to see what, if anything, was read out in the location where the anomaly was observed. This would be a simple and direct way to let regular, ordinary observers get information about alien anomalies they have observed. The data collected in this way could also provide useful information for shaping future research activities.

Measurements of alien anomalies in this way are extremely important. In an essay entitled, "The Function of Measurement in Modern Physical Science" (1961), Kuhn explains the critical role that measurement plays in defining the actual fit between scientific theory and the real world. In most cases, Kuhn tells us, theory guides the act of measurement and makes the real numbers happen. "[I]n scientific practice," Kuhn writes, "as seen through the journal literature, the scientist often seems rather to be struggling with facts, trying to force them into conformity with a theory he does not doubt. Quantitative facts cease to seem simply 'the given'. They must be fought for and with, and in this fight the theory with which they are to be compared proves the most potent weapon. Often scientists cannot get numbers that compare well with theory until they know what numbers they should be making nature yield." [79] As further explained by Kuhn,

> . . . new laws of nature are so very seldom discovered simply by inspecting the results of measurements made without advance knowledge of those laws. Because most scientific laws have so few quantitative points of contact with nature, because investigations of those contact points usually demand such laborious instrumentation and approximation, and because nature itself needs to be forced to yield the appropriate results, the route from theory or law to measurement can almost never be travelled backward. Numbers gathered without some knowledge of the regularity to be expected almost never speak for themselves. Almost certainly they remain just numbers.

---

[79] In the only peer-reviewed paper published thus far in this field, *Initial Results From the First Field Expedition of UAPx to Study Unidentified Anomalous Phenomena*, the team at UAPx was unable to derive conclusions as to the precise nature of the alien anomalies they studied, but they did suggest a very useful set of conventions to guide future research. This is precisely the type of "normal" scientific activity that Kuhn's paradigms suggest is necessary in these cases.

According to Kuhn, the scientist must always bring to bear her imagination, intuition, and vigilance in advance of the endeavor, not afterwards. But alien anomalies may upend the general nature of the rule because of the presence of possible real aliens causing the anomalies. If a scientist studies an alien anomaly, and a real alien is causing it, then the scientist is not actually studying "nature"; instead, she is studying the work of a superior living being. Embedded in the measurements are the fingerprints of the alien technology, which is only possible because the alien has already done the work of bringing alien versions of "imagination, intuition, and insight" into the formation of the technology. So in the case of alien anomalies, if we happen to be lucky enough to study one that is caused by a real alien presence, the measurements will have an embedded meaning, unlike the conventional approach where the scientist is simply studying the mysterious, mechanistic, depersonalized subject of Nature. Put differently, if a scientist studies an alien anomaly and has a real chance to measure it while it is active, she is not simply studying a phenomenon in the mechanistic sense, but she is also studying the work of an *active alien self*.

In addition to considering these changes to practice, science may need to reform the scientific discourse itself as it relates to alien anomalies (for regular, "non-numinous" objects of science, the conventional model will still work). Because aliens are inherently numinous objects, there may need to be some artificial restraints put in place to counteract the irrational forces they activate in the human mind. In practical terms this means, for instance, that if someone is going to question the plausibility of Loeb's argument for Oumuamua being an alien object of some sort, they be firmly held to account in all aspects of their reasoning and not be allowed to easily escape the difficult questions presented by the anomaly itself. Imposing a new form of regularity on scientific discourse about alien anomalies, borrowing elements from judicial discourse when appropriate, may be necessary in forming and shaping the discourse in order to develop a body of genuine scientific knowledge, while still allowing people to continue to disagree with one another and state their own views.

Secondly, it would be helpful to have the organs of government involved as team players in this alien enterprise, instead of always acting like hostile forces. Thus, another aspect of the vision of disclosure would include government reform and transparency on this issue in order to ensure that the America government at least is acting in a way that is consistent with the values of the First Amendment. As the Supreme Court stated in the case of *Thornhill v. Alabama*, 310 U.S. 88, 101-02 (1940),

> The freedom of speech and of the press guaranteed by the Constitution embraces at the least the liberty to discuss publicly and truthfully all matters of public concern without previous restraint or fear of subsequent punishment. The exigencies of the colonial period and the efforts to secure freedom from oppressive administration developed a broadened conception of these liberties as adequate to supply the public need for information and education with respect to the significant issues of the times. The Continental Congress in its letter sent to the Inhabitants of Quebec (October 26, 1774) referred to the "five great rights" and said: "The last right we shall mention, regards the freedom of the press. The importance of this consists, besides the advancement of truth, science, morality, and arts in general, in its diffusion of liberal sentiments on the administration of Government, its ready communication of thoughts between subjects, and its consequential promotion of union among them, whereby oppressive officers are ashamed or intimidated, into more honourable and just modes of conducting affairs." Journal of the Continental Congress, 1904 ed., vol. I, pp. 104, 108. Freedom of discussion, if it would fulfill its historic function in this nation, must embrace all issues about which information is needed or appropriate to enable the members of society to cope with the exigencies of their period.

Clearly these values are directly implicated in the study of alien anomalies.

Lastly, new social networks may be needed within which regular people can discuss their experiences with alien anomalies, or just share their interest in the topic, in a non-judgmental, anonymous way. Regardless of whether the anomalous experience was "real aliens" or not, the experience felt valid to the person and needs to be socialized somehow; if not, the experience turns into a neurotic repression which can only have negative effects.

In this connection, think back to the example of Ezekiel. Does it matter whether he really saw Yahweh or an alien space craft? To some it might, but to a modern scientific curious mind the important part is to be able to discuss and process the experience and have the experience respectfully examined. Only when such a process is allowed can the truth come forward. Perhaps it would turn out that a modern-day Ezekiel might realize that he didn't really see an alien or God, but had a mental health episode of some sort, prompting him to get help; or, conversely, by sharing the experience with others, it might become more real and significant to him. Regardless, sharing the experience, having it heard, and having some form of social integration and feedback is what counts; otherwise, the experience becomes entirely idiopathic and has no greater social significance. After all, if Ezekiel had felt too embarrassed or ashamed to share what he had witnessed with the people of his time, we would not have had his story to look at and consider today. These human stories are valuable and interesting, they are a part of us regardless of their source and should not childishly ridiculed or suppressed.

Substituting a realistic vision of disclosure for the current obsessions with revelatory forms of disclosure is one possible way to approach the conundrums presented by alien anomalies. Because people have been understandably obsessed with "revelations of disclosure" to date, we don't know much more about alien anomalies today than we did in 1958 in spite of our many technological advances in other areas of science. And while the gatekeepers of disclosure keep telling people to wait year after year after year for the "big reveal", it never seems to actually happen. This is a sign of a lack of a scientifically mature form of discourse.

7.3 *Casting for aliens*

We've talked about the relationships between humans, birds, and bees, but the relationship between humans and fish may provide a more fitting description as to how conventional human reality can relate to aliens at the present time. What we are talking about here is an example of an "impersonal" relationship, as opposed to the intimate relations that can form during the course of alien experiences.

The conventional point of view on human-to-alien contact is that it must be a revelatory event in which universal understanding and knowledge is transmitted to us in a divine flash (again, an experience that partakes of supernatural communication). While some people claim to have had these experiences during their alien encounters, this is not how the phenomena presents on a large-scale. Instead, these alien anomalies appear to us much more like fish do. We know they are there, somewhere, lurking beneath the surface, appearing from time to time and every once in a while we catch a more direct interaction and experience with their presence.

Humans and fish are both familiar and alien to each other at the same time. There is a very strange and limited form of “communication” between fish and humans, but it does exist. Through a complex, well developed system of knowledge that reflects itself in a wide range of fishing techniques, humans have learned how to directly connect with fish to get them to strike at a lure, bait, fly, nymph, streamer, or other human-created form of object. We can easily imagine that the earliest forms of fishing were just attempts to corral fish into a particular location so that a human could swoop down from an alien plane and grab them. Later, this developed into the use of nets and live baits. But at a certain point, it developed into artificial means as well. These artificial means were designed in a certain way to mimic the understanding of fish, deceiving them into believing that they were food, when they were not. This is a form of intimate, “personal” communication between the fish and the human across their alien orders.

When a fish strikes an artificial human device, a connection of communication has been made. The amount of information is, of course, limited, but it is the reflection of the work of a paradigm. Within that limited dimension of the fish strike, something real has happened, as connection has been made as a result of the conscious activity of the human.

This analogy breaks down, of course, because unlike fish, aliens are rationally presumed to be operating from a higher order of knowledge. We assume that this higher order easily encompasses our order, but this may not be the case. As we have seen in this work, it is arbitrary to assume that our understanding of the universe is universal. Aliens that have developed from radically different social, biological, or planetary platforms may not be relatable to us in the ways that we have imagined. We may be able to connect with them, but only in very limited dimensions, just like we can connect with fish (or, in some cases, if these aliens arise from biological lineages that are sympathetic to our own, deeper levels of personal communication may be possible). But still, even when a connection is limited, we learn something. Generations of fishers have developed increasingly sophisticated methods and means of catching fish, allowing them success in a wide range of environments. Through the act of catching fish, something is learned. The expectations here are, of course, more modest than the conventional view of human to alien connection, which always wants to turn it into an act of divine revelation. But if we are pursuing this subject from the standpoint of genuine scientific process, we have no need for such an event. We will be satisfied if we simply continue to learn and expand our knowledge. Also, like fishing, we will understand that no matter how well prepared we are, sometimes we won't catch anything at all. But still, we will try and be grateful, surprised, and astonished for any new pieces of information we glean. We will patiently add these experiences to our growing pool of knowledge, building our alien paradigms through steadfast and persistent efforts. Maybe we won't have a divine revelation that solves all of our problems, but we will still be learning something about ourselves and our world and increasing our pools of knowledge.[80]

---

[80] For example, an approach to aliens that mirrors the approach we use in fishing could result in the patient development of a catalogue of "signatures" for alien anomalies. In the field of fly fishing, for instance, fly fishers have developed an immense number of fly, nymph, and streamer patterns that vary in color, size, shape, and complexity. These can be applied in innumerable situations. Just go into any fly-fishing shop in a Western state and you can see what we mean. We could envision a situation where "catching aliens" (which here means a direct physical detection coupled with some kind of personal report) would give us more detailed information about their presentation. Going back to the Pascagoula Encounter for a minute, we can imagine how a robust surveillance system in that area during that time would have likely yielded a tremendous amount of physical data about the status of the aerospace. It would have allowed us to see how the anomalies moved, how they loitered, where they went to, when they returned, the characteristics of the way in which they displaced matter, heat, and air, and other characteristics as well. These, combined with what we learned from the human reports, could be assembled into some sort of signature, some sort of pattern. We may not end up understanding the "why" of any of it, but with this signature in hand, we might see it repeat in another encounter. Simply establishing

Instead of seeking disclosure, we could focus simply on developing conscious contact with aliens (if one assumes, of course, that some of what is occurring is possibly caused by real aliens). Think back again to the example of the honey guide bird. This is an anomalous instance where humans and birds can engage in direct communication. Of course, this type of communication or contact does not represent the full mode of conscious communication that exists between humans. For example, we cannot communicate complex paradigms and scientific ideas to birds. It would be insane to suggest that such a thing could happen. However, we are able to communicate some forms of rudimentary geospatial location between the bird and human in that particular case.

Applying this to the subject of human to alien contact, we would be to aliens what the honeyguide birds would be to us; thus, it would be arrogant for us and foolish to think that we could engage in a full range of communication with alien beings. In previous sections, we set forth all the reasons why this is likely the case. These include the fact that some aliens may have developed upon non-simian lines and have been in existence for quite some time and could be operating on paradigms which are as "advanced" in relation to ours us, as many of our paradigms would be in relation to people prior to the era of Copernicus or during the age of alchemy. In this type of situation, there is so much that simply cannot be communicated or understood. This doesn't mean, of course that we can't have any communication whatsoever with people from those prior eras of thought. We could, for instance, probably communicate with them regarding some basic matters of human life. However, it

---

some sort of "identity" among alien anomalies (i.e., showing that the same type of alien anomaly has happened on more than one occasion) is a form of real knowledge that is within our grasp. It would not reveal the secrets of the universe to us any more than a fisher who learns to catch a highly educated trout believes that they fully understand fish, but it is a form of meaningful activity in and of itself. It may be a more modest goal, but because it is attainable within the realm of scientific processes itself, it is certainly worth pursuing—at least it would be a much more valuable use of our time and resources than continuing to make more pseudo-scientific "alien entertainment" (i.e., documentaries) which don't lead to any new knowledge at all and just stir up emotions.

would be irrational for us to assume that we could have a complicated lecture in nuclear science or biochemistry with someone from the 16th century; so too, we must assume, would a "real alien" be limited in its communication with us. It is only our unlimited human arrogance which makes us falsely believe we could be on par with beings of such cosmic depth and complexity.

As we have seen in the discussion of Anaximander, advances of knowledge in the field of scientific thought often occur by drawing analogies between different orders of activity. Let us compare this directly to the challenge of catching fish. In the case of fishing, the fisher is tasked with figuring out what kind of lure, fly, bait, nymph, or streamer would be sufficiently similar to food a fish might be interested in to stimulate a bite reflex in the fish. This is an example of the presentation of knowledge across orders that are alien to one another; the intersection occurs when the fish takes a bite. The fish could not even have the merest inkling of the alien paradigms in which the fisher operates to design such products, yet if the fisher is successful, the bite occurs.

The fisher does not expect to communicate the full range of his or her knowledge to the fish, but merely one dimension of it. Just enough to get a bite, which presents a meaningful data point. Over the centuries that fishing has existed, the range of human knowledge on this subject has evolved with increasing complexity. Here too, in the case of alien anomalies, what we could be seeking, instead of a full range of alien knowledge, is just a "bite" of the alien anomaly, just enough data to establish the presence of something real. Focusing on getting these strikes, instances where our systems of observation connect with the presence of alien anomalies in a real way, is a very meaningful and probable form of productive activity, whereas developing the range of knowledge sufficient to understand much of what the alien paradigm itself may be seems out of the picture at the present time. In a sense, this activity might be said to be the obverse of fishing. Here, we may put a tremendous amount of activity into establishing the barest forms of knowledge. Just as the alchemists spent centuries operating in the dark, attempting to uncover the most basic elements of chemical processes whose contours they simply could not understand until the work of Boyle and Dalton ripened under Mendeleev, so here too perhaps we can see that the more modest expectations of knowledge actually represent its highest and most productive forms; we therefore accept the limits of our present knowledge, viewing ourselves as mere participants in a scientific journey whose course will exceed the limit of any individual lifetime.

Under this reasoning, we can say that part of the challenge that we face now in articulating real and genuine alien paradigms is to have realistic expectations and assessments about the types of things and ways in which we could actually communicate with aliens. A logical starting point to begin to build such realistic paradigms is to think back again to the case of the beaver. We begin such researches by focusing our efforts squarely on the externally observed features of these aliens, and refrain from engaging in unsupported and arbitrary speculations about their motivations or the ultimate details of their alien paradigms. But we can still look at external manifestations, which could allow us to make deductions about what types of paradigms could be in operation. We can perceive certain “alien gestures” and use them as starting points for further inquiry. We don’t have to know all the details of their paradigms in order to begin engaging in some learning, just as people in the Age of Alchemy did not have to understand the Periodic Table in order to begin investigating the world in a way which would one day lead to the formation of such a paradigm. But we today have a level of knowledge about how paradigms work that the alchemists and chemists did not due to our analysis of the record of human thought and activity (and in particular, the Kuhnian analysis of knowledge over time). We are suggesting that this additional level of knowledge must be applied to the study of alien anomalies and that without this knowledge, we will remain stuck. This is why we say that in the real arena of alien paradigms, we do not pursue the goal of disclosure because it is revelatory in nature, as opposed to scientific and incremental.

As contended throughout this work, a more productive approach to aliens at present is to treat them as *inferred objects* as opposed to “real” ones. This means we should accord aliens at the present time the same status that the atom had before Rutherford, the ether had before Einstein, and that dark matter has today. In each of these cases, there is compelling direct and circumstantial evidence upon with a rational mind can posit the object’s existence; however, there is a meaningful and substantial difference between an object’s status as being real versus inferred. When handling an object as “inferred” we are saying that we believe it should exist, but that we have not yet unearthed enough direct evidence to state that it does. The object is real in our minds, and we treat it as real in our minds, but we don't know yet if it will be real in the same way that other normal objects tend to be. To arrive at a proper result, we are willing to maintain our judgment in a purely provisional status, holding open the possibility that we may be wrong after all, and that the inferred object may not exist. At the same time, by concluding that the object could be real, we are willing to risk our time, energy, commitment—and in some cases our personal reputations—to do the work that needs to be done in order to see if the object is real or not. When we handle it as real in our mind, we have a chance at actually finding it in the world someday, just as Mendeleyev "found" the table, (which was at first no more real than a dream), and provided it status in the real world (as noted, this process has happened in many other instances as well, including, most notably, the concept of human flight).

From another angle, one could say that inferred objects are a discrete conceptual tool that humans use to explore the sources of invisible or unknown causes in order to create the conditions needed to possibly find a cause that is real and workable. The characteristics and assumed details of the inferred object often determine what types of detection methods are developed and deployed. If the inferred object is framed clearly and carefully in the realm of concept, it becomes possible to start looking for it by measurement in the real world. At some point, these invisible orders may become "real" and generate a new place in the settled order of conventional human thought as discussed in **Section 2.1.** In modern scientific practice, actual discovery of the inferred object has often been the result of new measurement techniques and technologies that were determined to be necessary in order to locate the object. These techniques were developed after the inferred object was thought about in the right way; when it was thought about in the "wrong way", it wasn't found. The atom and the ether present us with two different ways in which the search for the inferred object resolves. In the case of the atom, it was discovered and opened a gateway into new realms of space; in the case of the ether, the enhanced and precise systems of measurement opened into the realm of relativity theory when Einstein became involved, but demonstrated that the inferred object was an illusion.

Treating aliens as inferred objects makes conceptual sense only up to a certain point, however, because unlike the atom, ether, or dark matter, aliens are presumed to be macroscopic, sentient beings with independent powers of thought and movement, and who presumably operate within orders of knowledge that we do not fully understand. On top of that, aliens are not to be found solely by looking in microscopes, particle accelerators, or other esoteric instruments of modern science; by definition, in fact, if aliens are real, they should be accessible in the conventional world of embodied human experience. And indeed, many people claim to have had such alien experiences, even though science has not generally confirmed their reality. This shouldn't surprise us, however; for if these aliens really are beings that operate in higher orders of physics and knowledge, it would take us some time before we could establish them normally. Just think back to the tic tac. If someone had placed the tic tac encounter in a science fiction movie, no one would possibly think it could be a real occurrence, because we have no way to explain it. The only reason we do think the tic tac is probably a real alien, is based on the circumstances of the human experience and the quantum of both direct and circumstantial evidence.

This, of course, is where things get weird and difficult to explain because very few of the alien experiences have the level of circumstantial trustworthiness shown by the tic tac and generally they seem to behave like supernatural beings. And, of course, it is very easy to just conclude that a person reporting such an experience is mad. This is why "Fermi's paradox" tortures people with the question as to why we don't see aliens hanging around in the regular world; if aliens are real, and if they are here, shouldn't we be seeing them quite easily and shouldn't they display a presence which is global, irrevocable, and unmistakable to all?[81] We could understand why other inferred objects, such as the atom, would be hidden and difficult to find. But in the case of aliens, we modern humans naturally believe (or, rather, naturally *want* to believe) that we shouldn't have to look around for real aliens at all; thus, if we find ourselves having to look for them, then they must not be real.

Of course, if we analyze this conundrum carefully, we can see that the problem here is not the aliens—it's us. We are the ones that

[81] After all, isn't this what today's humans would do if they had the power to travel across galactic ranges at will and visit more primitive species on other planets? From this, shouldn't we conclude that every other species in the universe would do the same thing? lol.

are posting certain characteristics for these inferred objects, not the universe itself. We have decided that aliens would want to be our friends and reach out to us if they arrived here. We are the ones that excluded, on an *a priori* basis entirely, the possibility that aliens might come to earth, see us, and have absolutely no interest in being a part of the human world at all. We are the ones who decided that aliens must be like supernatural beings. The universe didn't force us to take this position.

In any event, this is one side of the strange contradiction we face here; namely, aliens aren't behaving as we 21st century humans would expect them to and therefore, many conclude, they simply aren't here. On the other side, however, there are many who have seen one too many anomalies (either through first-hand observation or experience, or by seeing the evidence produced by others) to be able to comfortably write the subject off entirely.

In this work, a description has been presented for the kind of aerospace survey that is needed as well as rational arguments for why that survey is needed. However, it seems unlikely that the necessary survey will happen any time soon. At present, the overwhelming majority of attention and energy is being placed into the search for "disclosure", which is not a normal scientific mode of knowledge, but more akin to religious revelation. Because so much attention and energy is sucked into the search for disclosure, there is little to no energy for anything resembling a more normal scientific inquiry. Because such an effort will take time, effort, money and will not necessarily be sexy, stimulating, and spooky, most people don't have the patience to do it. They want to get the glowing, numinous "high" from connecting with aliens, or being seen by their peers as the celebrity gatekeepers of the mysterious, alien world. While there are some fledging scientific attempts in the right direction, their scale and scope is far too small to actually lead to scientifically meaningful results. Moreover, some of these projects, such as Professor Avi Loeb's *Galileo Project*, while well-meaning and important, are based on flawed and incomplete theoretical and methodological premises that are highly unlikely to lead to any real results.

While there are no indications that an appropriate and productive scientific survey will happen anytime soon, it appears inevitable that it will happen after the present fledgling efforts have failed and enough people are willing to take seriously an approach along the lines discussed in this work. However, it will likely be some time before that actually happens due to the numinous pull of alien anomalies on the human mind.[82] Simply put, too many people are far too comfortable pursuing the delusions of disclosure, and are unwilling to reexamine their thinking on this subject to find the productive way forward. In the meantime, does that mean we just write off alien experiences into the realm of speculative science fiction and ignore them? Do we wait until we have "proof" that the experiences are real?

We suggest that it is consistent with Kuhnian theory to at least explore the possibility of conscious contact with aliens at the present time, provided that one properly treats the matter as pre-scientific, anecdotal evidence. Disclosure, by contrast, makes no sense. One gateway into this exploration, for those who chose to engage in, is through the mysterious realm of alien experiences. Another is to pare down our expectations of physical detection to something as basic as getting a fish to strike a tasty looking lure or bait. These strikes communicate valuable information, even if we wish that it was possible to get much more. In time, we hope that we will.

7.4 Conclusion: Cocconi and Morrison's revolutionary vision

While significant aspects of Cocconi and Morrison's work rest on a series of arbitrary assumptions, at the end of the day, the core of their analysis remains valid and revolutionary in nature. Because of their work, intelligent aliens are no longer just objects of religion or science fiction fantasy but are instead cognizable categories of causation within the realm of real scientific thought. As a result of their work, no one today should be ashamed to say that a particular anomaly could be caused by an alien, if the direct and circumstantial evidence supports such a conclusion. As stated by

---

[82] There is a certain type of patient, sober attitude that must be developed with regard to alien anomalies that is strongly resisted by these objects numinous effects. Think again to Von Frisch and his study of bees, and the patient, disciplined approach he took (and not the version of Von Frisch who is tripping on LSD). We must learn to handle alien anomalies in a similar, more "normal" way, in order for a real, non-delusional relationship to emerge. But so long as the aliens are treated as mystical, quasi-religious beings that excite unnatural thoughts and emotions, it remains exceedingly difficult to handle them as such.

Cocconi and Morrison at the end of *Searching for Interstellar Communications*,

> The reader may seek to consign these speculations wholly to the domain of science fiction. We submit, rather, that the foregoing line of argument demonstrates the presence of interstellar signals is entirely consistent with all we now know, and that if signals are present the means of detecting them is now at hand. Few will deny the profound importance, practical and philosophical, which the detection of interstellar communications would have. We therefore feel that a discriminating search for signals deserves a considerable effort. The probability of success is difficult to estimate; but if we never search, the chance of success is zero.

Cocconi and Morrison opened the door to a consideration of intelligent alien beings as real possibilities in the human world. They inferred the existence of intelligent alien beings in the universe based on sources of terrestrial and cosmological knowledge, as we discussed in **Section 3.3.1.** But as we noted in our critique of the SETI paradigm, Cocconi and Morrison also made a number of assumptions that are likely arbitrary and in error. Of these assumptions, the most problematic one is the notion that intelligent alien beings have not yet visited earth and can only be found in deep space. *See* **Sections 3.3.3.1 & 3.6.** This assumption is highly problematic because it inhibits investigations into the occurrence of alien anomalies in earth's aerospace.

Today, we see that this assumption should be called into question because alien anomalies are appearing in the human world in a non-visionary, scientific manner, as discussed in **Section 2.11.** We can also look back upon the record of human experience and see that these alien anomalies have always been a feature of our world. **Section 2.10**. What is changing today is how we can view them. In prior eras, these anomalies had to be either denied, ignored, or classified as being the result of the activities of supernatural beings. But today we can rationally consider that some of these anomalies could be the result of real alien activities occurring within the aerospace of this planet. **Section 1.7.**

If any alien anomalies are actually being caused by "real aliens" passing through our aerospace, then a study of these anomalies would necessarily open new fields of knowledge for us. This is at least what the scientific record establishes, as we noted in our discussion of anomalies and paradigm shifts. *See* **Sections 2.13, 2.16, and 5.4.** If we wish to explore this new path of knowledge, there are several steps we must take. To begin, we must situate these anomalies in their proper place within the conventional scientific order itself. This means that we must classify them as "pre-scientific phenomena" as discussed in **Sections 5.2** and **5.3.** Second, we must think about these anomalies in the clearest and most precise terms within the field of thought itself. This means that we must eschew fantasy, mystical, or spiritual speculation about the ultimate causes for any of these anomalies and study these anomalies like we would study bees and beavers. Instead of getting waylaid by mystical distractions (which are coming from us, not the anomalies themselves), we must train our attention on gaining as much information and depth of knowledge about the presentation of the anomalies in their own right, borrowing a page from the transformation of clinical medicine, as discussed in **Section 4.4.**

As we noted in **Section 2.17**, when we admit alien anomalies into scientific study, the knowledge potentials become meaningful. This requires us to treat the occurrence of alien anomalies as more than a mere historical project (i.e., determining what might have happened in the past), but rather on actively studying them as they happen. This, of course, is an extremely daunting challenge because we have no idea when, why, how, or where these anomalies will present themselves. But although daunting, time-consuming, and expensive, at bottom this is merely a *technical challenge*, and not a conceptual one. As shown by our discussion of the tic tac anomaly in **Section 2.11.1**, we have the capability to gather quite a bit of information about alien anomalies if we really want to.

The real conceptual challenge is generating the collective interest and effort into putting the same amount of time, energy, and resources into studying alien anomalies as we did studying atoms, subatomic particles, chemical processes, or gravitational waves—all of which began their journey from the invisible realm to the conventional one as *inferred objects*. One of the main points of *Alien Paradigms* is to overcome this resistance with a straightforward, plain vanilla, rational argument. Our position is simple: just like the Copernican Revolution opened unimaginable horizons of knowledge that couldn't be understood in the mid 16th

century, so too will the study of alien anomalies likely lead to the same outcome for our species, radically and irrevocably changing the nature of human reality for good.

In this Conclusion section, we have been as specific as possible in suggesting what type of knowledge might actually be available to us at the present time. The first order of business, of course, is just establishing that these anomalies are real and what precisely they seem to be doing in the aerospace (even if we don't know why they are doing it). Once that happens, the scientific imagination of a new generation of thinkers will be unleashed because they will understand new possibilities of physical motion which were never contemplated before. Powered by this new scientific faith, and a new vision of what is possible, they will likely be inspired to measure, detect, and invent systems to make those possibilities available for us.

Beyond the physical knowledge we might gain, the tantalizing and bizarre possibility of communicating directly with an intelligent alien species beckons. One could argue that actually happened in the case of the tic tac at some level, and people such as Charlie Hickson and Whitley Streiber claim to have experienced something even more direct; again, establishing the reality of these occurrences in conventional systems of thought is critical, and science provides the methodology to do so. Once that happens, more "regular" people will be open to these possibilities, and they will likely become a more normal feature of the human world, changing our future in ways we can't even imagine today.

Partial Table of Citations

INTRODUCTION

Thomas Kuhn, *The Copernican Revolution: Planetary Astronomy in the Development of Western Thought* (1959), pages 186-187, 189-192.

ALIEN ANOMALIES

David Bohm, *On Creativity* (1996), pages 15, 54-55, 120.

Donald Griffin, *Animal Minds: Beyond Cognition to Consciousness* (2001), pages 99-112.

Richard Dawkins and Yan Wong, *The Ancestor's Tale: A Pilgrimage to the Dawn of Evolution* (2016), pages 223, 227-228.

Kuhn, *The Copernican Revolution*, pages 5-6, 26, 137-38.

Thomas Kuhn, *The Structure of Scientific Revolutions* (1962), pages 5-6, 14-18, 33, 57-59.

C.G. Jung, *Flying Saucers: A Modern Myth of Things Seen in the Skies* (1958), pages 8-9, 10-11, 19, 23, 95-97, 109-110.

Megan Garber, "The Man Who Introduced the World to Flying Saucers", *The Atlantic*, June 15, 2014

Joan Bird, *Montana UFO's and Extraterrestrials* (2013), pages 15-76.

David Bohm, *Causality and Change in Modern Physics* (1957), page 99.

Thomas Kuhn, *The Essential Tension: Selected Studies in Scientific Tradition and Change* (1977), pages 31-65, 66-104, 209.

SAFE DISTANCE PARADIGM

Walter Isaacson, *Einstein: His Life and Universe* (2007), pages 142-44,149-153.

David Bohm, *Thought as a System* (1994), page 101.

Kuhn, *The Structure of Scientific Revolutions*, page 25.

Dawkins and Wong, *Ancestor's Tale*, pages 6, 28-29, 48-49, 257-58, 283-87, 291-94, 309-312, 351-52, 494, 501, 504-05, 667.

David Bohm, *The Special Theory of Relativity* (1965), pages 58-72, 256-57.

Roger Penrose, *The Road to Reality: A Complete Guide to the Laws of the Universe* (2004), pages 13, 16-17, 28-31, 34-35, 40-41, 43, 59-60.

Bohm, *Causality and Chance in Modern Physics*, page 125.

Griffin, *Animals Minds*, pages 93-99, 113, 174-180, 190-211, 228-29.

SCIENTIFIC CONVERSION PROBLEMS

Kuhn, *Structure of Scientific Revolutions*, pages 4-5, 5-17, 19, 23, 39, 53-56, 65, 97, 119-125, 202.

Bohm, *On Creativity*, pages 6-7, 19-20, 54.

Kuhn, *Essential Tension*, pages 20, 118-19.

Michel Foucault, *The Birth of the Clinic: An Archaeology of Medical Perception* (1962), xi-x, xvii, xix, 4, 7-9, 107, 169, 196.

Kuhn, *Special Relativity*, page xx.

Avi Loeb, *Extraterrestrial: The First Signs of Intelligent Life Beyond Earth* (2021), pages 1, 5, 30, 31, 34-43, 44, 61, 80.

Jung, *Flying Saucers*, page 34.

ALIEN PARADIGMS

David Shapiro, *Psychotherapy of Neurotic Character* (1989), pages 3, 10.

Jung, *Flying Saucers*, pages 16-17, 22-23, 108-110,

Thomas Ligotti, *The Conspiracy Against the Human Race* (2010), xx, xxi, 39, 41, 42, 44, 80, 133

Edward O Wilson, *The Meaning of Human Existence* (2014), pages 32, 42-43, 110-122

Bohm, *On Creativity*, pages 2-3

Kuhn, *Structure of Scientific Revolutions*, pages. 202-3.

CONCLUSION

Kuhn, *Essential Tension*, pages 193, 197-98.

Made in the USA
Monee, IL
04 November 2026

67379695R00184